Social*Text* **150**

Urban Climate Insurgency

Edited by Ashley Dawson, Marco Armiero, Ethemcan Turhan, and Roberta Biasillo

Urban Climate Insurgency

An Introduction

Ashley Dawson, Marco Armiero, Ethemcan Turhan, and Roberta Biasillo

On December 1, 2020, Indian police arrested Bilkis Dadi, one of the most prominent activists of the women-led Shaheen Bagh protests against the Bharatiya Janata Party's (BJP's) exclusionary citizenship act, on her way to join a massive encampment of farmers on the border of the nation's capital city. "We are daughters of farmers," Dadi said, and therefore "we will go to support the farmers in the protest today."[1] Farmers, who constitute more than 50 percent of India's population, have been pushed beyond the breaking point. After months spent camping in the cold and rain on the outskirts of Delhi in an effort to get the government of Narendra Modi to repeal a trio of laws intended to privatize the agricultural sector, they decided to take their protest into the center of the country's capital city. On January 26, 2021, on the day India celebrates the anniversary of its inception as a constitutional republic, more than 150,000 tractors and hundreds of thousands more protesters on foot set out from the farmers' encampments on the fringes of the city for a peaceful march toward the nation's centers of power. Modi's efforts to stop the protest using legal injunctions had failed, but the leaders of farmer unions had agreed to keep the protesters on police-approved routes through the city outskirts—ensuring that the protest would not disrupt official celebrations. The farmers were having none of it. At the city's border with the village of Ghazipur, site of one of the farmer encampments, tractors pushed aside a shipping container placed on the road by the Delhi police.[2] Elsewhere, police responded with thick clouds of tear gas and baton charges as farmers tried to march off the capital's ring road and push toward the city center. Delhi police commanders placed officers with assault rifles across key routes

Social Text 150 · Vol. 40, No. 1 · March 2022
DOI 10.1215/01642472-9495075

into the city but the farmers refused to be kept out of the symbolic heart of the city. By noon, farmers had breached the Red Fort, the iconic palace that once served as the residence of India's Mughal emperors.

The day's events in Delhi, and the months of peaceful protests on the city's outskirts that preceded them, should be seen as the latest instance of an increasingly important global phenomenon that we call the *urban climate insurgency*. With this term we refer to the ensemble of grassroots initiatives that aim to tackle climate change from a radical point of view and take the city as the primary locus of action in doing so. In our vocabulary, *insurgent* does not imply violence but rather refers to the radical rejection of the current socio-ecological system. Urban climate insurgency does not follow the rules of the game; it does not legitimize the current climate regime through the paraphernalia of participatory tools that are designed to anesthetize anger and social mobilization. Employing Salvatore Paolo De Rosa's words in his article for this special issue, climate insurgency rebels against "the current post-political condition that attempts to foreclose politicization and evacuate dissent through apparent participation and technocratic expertise in the context of a non-disputed market-based socio-economic organization." It is insurgent because it clashes with mainstream climate policies, acknowledging that the climate crisis is not a mistake of the system but is the evidence that the system is deeply rotten and must be changed. Mobilizing farmers and their allies by the hundreds of thousands, the protests in India are the biggest civil society protests in the world, but they are far from being the only example of contemporary urban climate insurgency. In order to understand the full extent of the insurgency, it is important to understand that the climate emergency[3] forms a political unconscious that is a constitutive feature of all contemporary public events. This means that urban climate insurgency should be seen as a spectrum, inclusive of uprisings that explicitly challenge planetary ecocide, on the one end, but extending all the way to protests that target austerity policies that make the urban condition yet more precarious, on the other end. As Marco Armiero and AbdouMaliq Simone and Solomon Benjamin demonstrate in their articles, the urban climate insurgency should, in other words, have been seen as an inherently intersectional form of mobilization.

Consider the Indian farmer protests as an example of this intersectional quality of the urban climate insurgency. The overt intent of the farmers' demonstrations was to agitate for repeal of a set of laws rammed through the Indian Parliament in September 2020 by the right-wing BJP regime. These laws aim to dismantle the independence-era system of government regulation that assures farmers are paid a "minimum support price" (MSP) for their crops, thereby protecting the farmers, as well as the general public, from the vicissitudes of the free market, both local and global.[4] Although the government markets buy only one-third of the crops produced in India, the MSP acts as a benchmark figure that impacts negotiations for prices throughout the agricultural sector. Under the new laws, farmers would be allowed to sell their produce directly to private buyers rather than at the state-regulated marketplace, to enter into legal contracts with private companies for their crops before harvest, and to hoard grain until prices increase. Farmers have been outraged by the swift passage of the laws, arguing that they would allow large corporations to displace the small traders who currently dominate the government-operated marketplaces, drastically shifting the balance of power and curtailing farmers' capacity to negotiate fair prices for their crops. Since more than 50 percent of the Indian population works in the agricultural sector, the privatization of agricultural markets has stark implications for the country as a whole.

Behind this battle over the privatization of agriculture lies the climate emergency. In India over the last half century, extreme weather events, particularly extreme precipitation, have tripled in number. Rainfall has grown less frequent, but when rains come they are torrential and frequently destroy crops. Since roughly half of India's farmers are too poor to afford irrigation and therefore depend on rain-fed agriculture, increasing weather extremes translate into intense economic vulnerability. But even more well-off farmers are in crisis, since they typically attempt to cope with weather extremes by increasing the use of expensive inputs like chemical fertilizers, pesticides, and irrigation, which degrades the quality of soil and lowers groundwater levels.[5] The upshot is decreasing productivity and growing debt levels, as farmers spend more to grow less. Their efforts to switch to less risky crops ironically lead to overproduction and the crashing of prices for both major and minor crops. As a result of this vicious economic cycle, Indian agriculture exhibits the same instability and extremes as the weather conditions imposed by the climate emergency. Caught in this climate vice, farmers have demanded increases in the minimum support price set by the government, but the Modi regime has instead decided that it is unsustainable for the government to continue to subsidize an uncompetitive sector. Cut loose by the state, farmers face impossible choices. Increasing numbers have been taking their own lives. The climate insurgency in Delhi is a defiant antithesis to such gestures of hopelessness.

The protests of Indian farmers in Delhi conform closely to the paradigm of "environmentalism of the poor" laid out by Ramachandra Guha and Joan Martínez Alier in their seminal book *Varieties of Environmentalism*. In this work, Guha and Martínez Alier argue that environmental protests in India pit "ecosystem people" (communities that depend heavily on local natural resources) against "omnivores" (individuals and communities with the social power to control and use far-flung resources).[6] For Guha and Martínez Alier, the history of "development" in India is the story of resource capture by omnivores, a process that is resisted by ecosystem people; examples they document include the struggle of tribal peoples against displacement caused by dams, and the resistance of peasants to diversion of forests and grazing lands to industrial uses. In the course of these struggles against displacement, ecosystem people typically deploy exceptional creative forms of direct-action protest that combine both a utilitarian and an expressive dimension. Protesters, that is, both try to shut down business as usual and to articulate why the social inequalities that subtend their dispossession are morally repugnant.[7] In the cases documented by Guha and Martínez Alier, this almost always involves invasion of the urban areas where the majority of omnivores live by large groups of rurally based ecosystem people. The encampments of Indian farmers outside Delhi and their marches into the city center on particularly symbolic days typifies the protest repertoire of ecosystem people.

The climatic and political contradictions playing out in India are typical of conditions affecting growing segments of the global population. Particularly among the peoples of the Global South, the worsening climate emergency is driving the proliferation and increasing political prominence of urban insurgencies around the world. These insurgencies sometimes are driven by protesters from rural areas but often are organized by large numbers of people who live in cities, many of whom are what Guha and Martínez Alier call "environmental refugees"—people pushed off their ancestral lands by forms of elite enclosure, as well as, increasingly, climate breakdown. From hunger riots in Santiago, Chile, in May 2020 to protests in Ecuador against rapid removal of oil subsidies to the detriment of the urban poor, from huge marches in the favelas of São Paulo to the state governor's palace by masses of people demanding economic support to anticurfew protests by slumdwellers in Nairobi, examples of urban uprisings over the last year alone are legion. As these examples suggest, the uprisings don't simply involve rural people, such as Indian farmers, bringing their plight to the attention of urban elites through various forms of direct action. The urban climate insurgency is also constituted by those urbanized masses whose life conditions are growing more precarious as a result of what we would characterize as a compound crisis: rampant economic and social inequality, waves of austerity and privatization, draconian state repression, and increasingly threatening climatic extremes such as life-threatening heat and dangerous flooding.[8]

The COVID-19 pandemic exacerbated these conditions, generating a crisis of brutal ferocity in which the most basic conditions of social reproduction were stripped from ever-larger numbers of people around the globe. The pandemic played out as an intensification of what Evan

Calder Williams has characterized as the mode of combined and uneven apocalypse of contemporary late capitalism, a shift from apocalypse as an instant, universal, one-off event to an archipelago of zones of infernal breakdown.[9] City dwellers, who by definition have been stripped of the relative autonomy from capitalist markets that characterizes the rural peasantry, proved particularly vulnerable to this combined and uneven apocalypse, as the thin margins that allowed them to purchase food were suddenly torn asunder, and bare survival became the horizon of everyday life. As Andreas Malm pointed out in his discussion of the Arab Spring, inability to access food has a famous capacity to radicalize: as access to the means of basic subsistence becomes a function of the unequal distribution of wealth, the ruling regime tends to be perceived less as a guarantee than as a threat to the bodily metabolism of its people.[10] Under such conditions, the ruling regime risks losing all legitimacy, while the people, faced with starvation, come to feel they have nothing to lose. Similarly, in this special issue, contributors such as Armiero, Lise Sedrez, and Roberta Biasillo argue that the unequal exposure to environmental threats—be it toxic contamination or floods—politicizes subaltern communities. If climate change is not the sole factor in sparking contemporary uprisings, it is increasingly a key ingredient in a planetary urban powder keg.

Given the intensifying prospects for combined and uneven apocalypse on an urban scale, we would argue that the city is an increasingly key site for insurrection—and, further, that the climate emergency is a constitutive feature of these insurrections, whether this is overtly declared or not. The importance of the urban scale in climate politics may seem counterintuitive given the long-standing opposition between cities and nature. Typically, cities and nature are perceived as geographical and cultural opposites, with cities seen as manufactured social creations while nature is seen as a space outside of human creation.[11] The climate emergency makes rejection of this stale dichotomy imperative. Cities are at the core of what radical environmental thinkers call the Capitalocene: the era when the capitalist system's frenetic drive toward incessant growth through the accumulation of capital has dramatically destabilized Earth systems.[12] According to the United Nations, cities are responsible for up to 75 percent of contemporary carbon emissions and 60 percent of resource use, with transport and buildings among the largest contributors to greenhouse gases.[13] Yet cities and the urban scale have been remarkably invisible in discussions of the climate emergency. Carbon emissions, for example, are most often reported on through national statistics, or in terms of per capita individual emissions that are themselves tabulated based on a nation-state framework. But since most of the world's cities are in low-lying coastal areas, urban populations are at the forefront of the climate emergency as seas rise and anthropogenic environmental disasters strike.

We do not mean, of course, to suggest that all contemporary climate-justice struggles have a predominantly urban dimension. The founding of the Sacred Stone Camp by Ladonna Bravebull Allard at the Standing Rock Reservation to resist the Dakota Access Pipeline catalyzed awareness and solidarity actions around the world in support of global Indigenous struggles against a new wave of extreme extraction. Movements of the world's peasants and smallholding farmers for food sovereignty and against capitalist agriculture are also key to addressing the climate emergency.[14] The resistance of communities of fisherfolk against the corporate enclosure of the oceanic commons is another example of nonurban struggles with a strong environmental dimension. And, finally, the Red Nation's *Red Deal* articulates demands for anticapitalism and decolonization that straddle urban and rural spaces.[15] We could enumerate other instances of such nonurban environmental conflicts; nonetheless, many contemporary social movements—even when they are not founded on explicit demands of a right to the city—have a primarily urban dimension.

In addition, the climate emergency manifests on urban terrain already fissured deeply by the unequal and exploitative geographies produced by racial capitalism. Precisely as described in this volume by Sedrez and Biasillo, Ashley Dawson and Macarena Gómez-Barris, and Simone and Benjamin, new forms of climate apartheid arise as extreme inequalities resulting from decades of redlining, gentrification, and forced displacement intensify vulnerabilities to climate-change-related crises such as searing urban heat or serial flooding, all while the global elites who are most responsible for carbon emissions retreat to safer sites.[16] Climate apartheid requires ideologies of violent othering.[17] As Naomi Klein puts it, "Fossil fuels require sacrifice zones: they always have. You can't have a system built on sacrificial places and sacrificial people unless intellectual theories that justify their sacrifice exist and persist."[18] As the climate emergency deepens, discourses used for centuries to dehumanize others while legitimating the extractive violence of capitalism, colonialism, and patriarchy increasingly target those whose lives are most menaced and who are displaced by fossil capitalism. Malthusian fears about impending waves of climate refugees become core doctrines of official state discourse. Climate apartheid thus builds on the suffusion of the public sphere by fear, a trend that is key to the success of contemporary extreme right-wing parties.[19] In the post-9/11 age, it has been easy to poison public discourse with fear and, of course, with its corollaries—that is, with a growing apparatus of control, surveillance, and meticulous classification. Xenophobia, homo- and transphobia, and blatant racism are probably the most dramatic effects of this political investment in fear. Walls at the borders and closed ports are the emblems of this phobic politics, transmitting in stark visual form the oppressive sense of terror that is permeating individual and collective lives in this new millennium. On a microscale, these infrastructures of fear appear to be embedded in the fabric of contemporary urban life with video surveillance, gated communities, armed guards, and de facto off-limits zones. The enduring resonance of this fear-mongering was evident when Donald Trump labeled cities such as Seattle and New York "anarchist jurisdictions" following the nonviolent Black Lives Matter protests of summer and autumn 2020.[20]

While manufacturing and intensifying anxieties against any form of diversity, authoritarian populists have deployed a strategy of disavowal for the climate emergency. From Donald Trump in the United States to Jair Bolsonaro in Brazil and Matteo Salvini in Italy, contemporary right-wing leaders have refused to acknowledge the gravity—or even the existence—of the climate emergency. Right-wing politicians have often ridiculed warnings about the climate crisis from scientists and environmentalists, blaming them for spreading unnecessary anxieties among people and prattling on about environmentalists' job-killing policies. Fear of migrants or nonbinary sexual identities but trust in pipelines, dams, and nuclear power plants—this is the odd reality of regimes that want people both scared and oblivious. After all, this is less contradictory than one might think: the (re)production of fears against those who "do not belong" is meant to channel frustrations and anger away from progressive forms of social struggles. In this special issue, both Sinan Erensü, Barış İne, and Yaşar Adnan Adanalı and Simone and Benjamin discuss the ways in which conservative articulations of belonging are intertwined with place-making processes and discourses. But within the far-right netherworld, this politics of fear is metastasizing into an overt eco-fascism, a toxic ideology that marries a Malthusian acknowledgment of environmental limits with violent white supremacist efforts to purify the nation.[21]

Given the prominence of this politics of fear,[22] and its successful capture of political power in many contemporary nations, it is not surprising that a global network of progressive cities has decided to coalesce under the banner "Fearless Cities." In the context of authoritarian populism and xenophobia, fearless cities stand for welcoming policies and solidarity toward migrants. However, their project is wider than that of advocacy on behalf of migrants. In the face of crisis and despair, fearless cities imagine themselves as spaces for nurturing "human rights, democracy, and the common good."[23] Hope over fear is the main message of the project, one that overlaps with the new municipalist agenda being articulated by social movements in cities around the globe. Although rooted in the European context, fearless cities are appearing everywhere. In June 2017, the political "confluence" Barcelona en Comú hosted the first international Fearless Cities summit, bringing together more than seven hundred officially registered participants from six continents. Regional Fearless Cities gatherings have occurred throughout 2018 in Warsaw, New York, Brussels, and Valparaiso. The 2019 global gathering demonstrated the breadth of municipalist movements: Zagreb je Naš in Croatia, Miasto Jes Nasze in Warsaw, Cooperation Richmond in California, the Umbrella Movement in Hong Kong, Ne Davimo Beograd in Belgrade, the Autonomous Government of Rojava, Cambiamo Messina dal Basso in Messina, Movimiento Autonomista in Valparaíso, Ciudad Futura in Rosario, Cooperation Jackson in Mississippi, Beirut Madinati from Lebanon, and a number of other Spanish cities such as Marea Atlantica in A Coruña and Zaragoza en Común in Zaragoza.[24] The most visible example of how this movement spun off outside Europe is that of the sanctuary cities in the United States—that is, municipalities, but also counties and states, that resisted Trump's anti-immigrant policies.[25] Tellingly, the map of the sanctuary cities overlaps, at least partially, with that of the cities that have declared their commitment to the Paris Agreement.

While this new municipalist project has been studied from many points of view, its environmental agenda remains quite unresearched.[26] This gap is even more relevant when one considers the growing interest in multilevel governance in climate change policies and, specifically, in the role of cities in the elaboration of mitigation and adaptation strategies. The smart city is perhaps the most popular version of this urban discourse on climate change. For philanthropic foundations with an urban focus such as C40,[27] a blend of technological innovations and entrepreneurship, combined with the scientification of political decision-making, seems to be the best solution to foster "greener or more efficient cities that are simultaneously engines for economic growth."[28] Scholars have demonstrated how "techno-ecological fixes" as tools to face climate crises are used to justify "crypto-colonialism."[29] First, they play into ongoing narratives of "green grabbing," where local claims to resources are liquidated for green investments. Second, technology perpetuates North-South trade and investment inequalities. And third, a new power asymmetry is enabled by the technology through data colonialism and surveillance capitalism. Massive Asian smart-city projects have been deemed the materialization of a "new colonialism" due to the undermining of local culture and overlooking communities.[30] The smart-city approach to climate resilience is never placeless and produces so-called franchise colonial dynamics. Indeed, such urban planning practices act as proxies for global financial and technological elites and almost always involve displacement of "informal" settlements.[31]

Other aspects of the urban efforts to tackle climate change are less explored, including how radical cities and social movements located therein are addressing climate change. Take Black Lives Matter, for instance. The movement might at first sight have little to do with the climate emergency, with policy documents such as the *Vision for Black Lives* focused on abolition of the carceral state and defunding of police forces.[32] Nonetheless, the Movement for Black Lives (M4BL) is highly aware of the intersectional character of the compound crisis. A recent M4BL teach-in, for example, highlights the "dual crisis" of climate change and COVID-19, and links this crisis to other aspects of the climate emergency that disproportionately affect Black people, including climate gentrification, environmental contamination, energy poverty, and displacement and forced migration, among other issues.[33] To challenge this compound crisis, M4BL calls for climate reparations and a Red, Black, and Green New Deal.

In this special issue we explore the articulation of radical climate change politics and materialization of climate justice at the city level, a topic finally receiving well-deserved attention.[34] As we have seen, our notion of urban climate insurgency refers to grassroots movements that arise in cities in the grip of the intensifying climate emergency, as well as in cities targeted by insurgent movements that spring up in rural areas thrown into crisis by climate change. We employ the city in a double-edged way, considering it as both an institutional and a spatial scale of analysis. In other words, we research municipal governments and their climate policies as well as grassroots initiatives practiced at the urban scale. As the majority of humanity has moved to cities over the last half century, questions of the urban environment and of urban governance have become increasingly central to social and political conflicts across the Global North and South. In addition, with right-wing populist governments in control of many national governments, struggles for economic equality, political inclusion, and climate justice have taken on increasing prominence on the urban scale.

What repertoire of protest characterizes the urban climate insurgency? Ramachandra Guha and Joan Martínez Alier have suggested that the "environmentalism of the poor" is characterized by strategies of direct action with three key goals: demonstrations of collective strength (e.g., mass demonstrations in key sites of power); disruptions of economic life (e.g., sit-down strikes); and acts intended to put moral pressure on specific powerful individuals or on the state as a whole (e.g., indefinite hunger strikes).[35] For Guha and Martínez Alier, these direct action tactics play out crucially in cities, sites of concentrated state power, media focus, and symbolic resonance. In the essays that follow, contributors track the protest strategies adopted by movements mobilizing in urban spaces over the last decade or so. From Tahrir Square and the Arab Spring, to the Puerta del Sol and Spain's Indignados, Ferguson and Black Lives Matter, the Gezi Park protests in Istanbul, Mong Kok and Hong Kong's Umbrella Revolution, and Zuccotti Park and Occupy Wall Street, the occupation of public spaces in cities has been a pivotal strategy for disparate movements over the last decade or so.

It could perhaps be said that the tactic of occupation is not that new. In the history of social movements and labor unions, the act of occupying has been both a defensive and a prefigurative strategy. Workers have occupied factories to resist their closure as well as to experiment with forms of self-management.[36] Peasants have occupied lands against big landowners, resisting enclosures and expropriation and sometimes trying to build alternative communities on such reappropriated land.[37] The occupation of buildings to fight eviction and secure affordable houses has also been part of a long-standing repertoire of social mobilization, sometimes producing new forms of living in common.[38] Sometimes the occupation of specific spaces and infrastructures has brought a halt to activities that were considered harmful to the environment and communities; these are the kind of stories told by De Rosa and Erensü et al. about Gothenburg and Istanbul.[39] In Italy, for instance, the word and practice of "occupazione" is strongly connected to the autonomist tradition and the radical non-mainstream left.[40] Over the last decade, this long-lasting and quite rich tradition of social movements squatting/taking control of buildings, factories, lands, and public spaces has often been reduced to the very specific experience of the Occupy movement, with its iconic camp in Wall Street. While offering a potentially rich venue for public dialogue and new forms of direct democracy, the Occupy experiments were also entrenched in a series of limitations that seemed to have reproduced racial, class, and gender inequalities. With all the best intentions, the slogans of the movement, "occupy" and "we are the 99 percent," revealed the weak engagement of the movement with the issues of coloniality and intersectional diversities, which, indeed, were not represented in most of those experiences.[41] As Joanne Barker notes, Indigenous dispossession was the historical precondition for Wall Street itself—a street with a wall built by the Dutch, in part, to keep the Lenape people out of their homeland in what became lower Manhattan.[42]

Evidently, most of these occupation experiments, especially those in public spaces, are fragile almost by definition and destined to be swept away by police repression. But if instead we consider the broader range of social movements that have arisen in the last decades, including the farmer protests in India we discussed at the outset, but also indigenous movements, Black Lives Matter, Fridays for Future, and *Ni una menos*, we can have a better sense of the contribution of grassroots mobilization to the dismantling of neoliberal orthodoxies, hegemonic for the previous three decades. Although not all directly engaged with climate change issues, those movements have developed forms of social networking that have proven to be a key element in mutual aid in the context of the climate emergency. Indeed, in 2012 the activists who were engaged in the Occupy Wall Street movement decided to react to the destruction brought to New

York City by Hurricane Sandy, relocating from the financial district to Brooklyn and Queens and in the process giving birth to the Occupy Sandy movement.[43] In this special issue Marco Armiero illustrates how Marxist autonomous radical activists in Naples, Italy, mobilized first around toxicity and environmental justice and then around climate change in a positive osmotic relationship between their political allegiances and new practices and concepts.

Often, those movements have recognized that "no is not enough" and that the evacuation of the space of institutionalized politics simply allows the right wing to cement its power. This has led to the growth of what we referred to above as the new municipalism: in cities such as Barcelona, Madrid, Berlin, Naples, and Jackson, Mississippi, social movements that began their lives fighting existing political institutions around issues such as the global housing crisis have now fought their way to political power. But this transition has not necessarily dulled their insurgent edge. In a case such as Barcelona en Comú, the social movement is attempting to transform the institutions of urban governance while holding onto its base in popular mobilization for participatory democracy.

The experiments initiated by the new municipalism, including cutting-edge plans for climate action, are reshaping people's sense of what is politically possible. The Fearless Cities network has helped to translate the sense of the possible sparked by this new municipalism beyond the traditional upscaling to the national level through the establishment of transnational connections between radical cities. One of the most prominent characteristics of this new municipalism is what might be called an "intersectional climate politics—"the recognition, that is, that struggles for adequate housing, for the right to mobility in the city, for food justice, and even for the defense of migrants are all related to the fight around the climate emergency. Cities built for the rich will never be just or sustainable. The demands articulated by the Movement for Black Lives are characteristic of this intersectional politics: for Black Lives Matter, demands for reparations include recognition of the systematic harm done to Black communities by environmental racism, food apartheid, housing discrimination, and racial capitalism more broadly. The Movement for Black Lives insists on divestment from "exploitative forces, including prisons, fossil fuels, police, surveillance, and exploitative corporations," and, instead of such oppression and extraction, "investments in Black communities, determined by Black communities."[44]

The frictions arising from the interactions between social inequalities and climate change politics have manifested themselves quite dramatically in a couple of iconic episodes of the most recent urban insurgencies. In Paris the so-called Yellow Vests Movement has mobilized against the introduction of a measure that was supposed to reduce CO_2 emissions by increasing the cost of fuel for cars. Some have argued that this was a demonstration of the lack of popular consensus on climate change policies. Instead, from an urban climate insurgency point of view, it was a demonstration only that there cannot be any climate consensus on policies securing class, race, and gender privileges under the umbrella of the climate emergency. Who will pay the bill for the ecological transition is not a secondary problem.[45] In London, the radical environmentalist movement Extinction Rebellion (XR) clashed with commuters while disrupting the public transportation services of the subway; the image of a white male activist hitting a Black commuter was a clear manifestation of some underlying blindness regarding race, class, and gender present in a movement like Extinction Rebellion.[46] The Brixton flowers incident was then a further demonstration of the blindness of the movement toward class and race inequalities.[47]

These various examples of urban climate insurgency make clear that climate policy is no longer the exclusive province of national governments, international agreements, and panels of experts. On top of this, "Alternative ideas of climate urbanism emerg[ing] from insurgent attempts at reclaiming urban commons and public space ideas . . . do not always match institutional efforts to shape urban environments," as Vanesa Castán Broto and Enora Robin argued recently.[48] Despite the fact that the city has traditionally been represented as the antithesis of nature, we argue in this special issue that it is currently a key terrain for environmental struggle.

The contributions in this special issue explore, expand, and ground questions such as, how do the understandings and experiments of grassroots urban movements struggling for climate justice differ from those of movements unfolding at other scales? In what ways does the urban scale help catalyze more-radical struggles? What are the obstacles that such movements encounter on the terrain of the global city? If, as Henri Lefebvre argued, urbanization is the key form of contemporary capitalism, to what extent do the peculiar characteristics of contemporary extreme cities help generate struggles both for survival in slum ecologies and for more just and sustainable cities to come?

Finally, we do not intend to assume that an urban politics for climate emergency is always a radical, egalitarian one. While contemporary right-wing populism and neofascist movements have certainly achieved bold yet fragile victories at national scales in places such as the United States, Hungary, Turkey, Brazil, and the Philippines, to take just a few examples, these movements also have gained purchase on urban terrain in various places. In Brazil, for example, the recent rise of the right began with public protests in cities such as Rio de Janeiro and São Paulo against failures and inefficiencies in public services like urban transit systems.

Very quickly, however, these popular protests were hijacked by the right into protests against political "corruption" that came to target the Workers' Party, leading to a constitutional coup against President Dilma Rousseff. What makes urban terrain fertile for such right-wing movements? As the climate emergency intensifies, leading to greater numbers of climate refugees (both within and across national borders), how can progressive urban movements counter and head off the unfolding of xenophobia within particular cities through expanding notions of urban citizenship?[49] What role can a progressive municipalist politics and antifa mobilization play in challenging the growth of such movements? These are some of the debates we would like to ground with the set of articles in this special issue complementing one another in more than one way.

Following this introduction, the special issue opens with the contribution of AbdouMaliq Simone and Solomon Benjamin on "majority urban politics" in times of climate emergencies with empirical attention to pro-poor politics in northern Jakarta and land occupations by low-income residents in Bangalore. In exploring the multilayered complexities in these two cases, Simone and Benjamin advance ideas on how to reformulate an urban life worth living in times of climate crisis. In doing so, they also extend the critique of "an imposition of fixity, measurability and transparency" of urban communities and challenges "involving a variety of institutional actors, from consulting firms, NGOs, university-based institutes and think tanks, and even activists."

The second contribution to this issue is the work of Ashley Dawson and Macarena Gómez-Barris on grassroots, postextractivist responses to energy colonialism, uneven burdens, and democratic control of energy. By focusing on the cases of UPROSE, Brooklyn's oldest Latinx community-based organization in the United States, and YASunidos in Ecuador, Dawson and Gómez-Barris propose the idea of energy states that hinges on a "radical interdependency that could be legislated by a substantially altered state apparatus during the implementation of a new energy paradigm." Positioning the focus succinctly on the contemporary Green New Deal discussions, this contribution calls for a postextractivist future in which land rights, social and multispecies justice, and antiracist social movements will be at center stage.

The third contribution comes from Marco Armiero, who takes us to the trash-filled streets of Naples, Italy, at the height of an urban garbage crisis. Focusing on how the embodied experience of contamination led to a political renewal, a novel force creating resisting communities and activist knowledge, Armiero's contribution shows the parallels and continuity between urban waste and urban climate-justice activism in a time when radical grassroots movements are increasingly waging intersectional struggles. In particular, Armiero explores how the category of

Biocide, elaborated by the antitoxic activists, has provided them with the analytical tools needed to embrace climate change from an insurgency perspective.

In the fourth contribution of the special issue, Lise Sedrez and Roberta Biasillo shift focus to the multispecies alliance for ecological justice in one of the most troubled favelas of Rio de Janeiro. Through their focus on landslides and other climate-related disasters in Morro da Babilônia, Sedrez and Biasillo problematize the sociopolitical and symbolic marginalization of favela dwellers and demonstrate how community-based groups in the favela rose to prominence as social and ecological protagonists. Using oral history, memories, and narratives, this piece argues that community-led reforestation initiatives contribute to "environmentalization of social struggles," in which the communities with high awareness of what is at stake take the lead in, against, and beyond the state and other nonstate actors. Their proposal of a multispecies alliance from the social margins of the city resonates with what David Pellow in his interview calls a "multispecies abolition democracy."

The fifth contribution of the special issue, by Sinan Erensü, Barış İne, and Yaşar Adnan Adanalı, critically analyzes the "urban greenery frenzy" of Recep Tayyip Erdoğan's authoritarian regime in İstanbul, Turkey. Situating the everyday politics of urban-environmental aesthetics in a broader political-economic and symbolic transformation, Erensü and colleagues trace how authoritarian populism creates its narratives and counter-narratives over urban green spaces and how movements with radical claims for the right to the city, such as those demonstrated in the Gezi Park uprising, fight back. The article concludes by calling for further attention on "opening up municipal practices to Gezi style radical experimentation and financial and political priorities."

The last contribution to this special issue, authored by Salvatore Paolo De Rosa, takes a situated urban political ecology approach in exploring the climate-justice direct-action coalition Fossilgasfällan in Sweden. Focusing on the debates on socio-ecological metabolism and urbanization, De Rosa offers "metabolic activism" understood as "grassroots eco-political engagements that aim to disrupt, block, occupy and ultimately transform capitalist-driven metabolic relations." This notion is elaborated further through the case study of Fossilgasfällan, an organization that generated the first-ever blockade of fossil fuel infrastructure through direct action in Gothenburg.

Intersectionality and antiracist struggles are at the core of the interview with David N. Pellow, curated by Armiero and De Rosa. In the interview, Pellow reflects on the Central Coast Climate Justice Network (C3JN) and its effort to bridge antiracist and environmentalist struggles.

The entire interview is focused on one of the key themes covered in this special issue: the unequal experience of climate change. Akin to what Simone and Benjamin argue in their article, Pellow also contrasts the white/Global North discourse on the "unprecedented existential crisis" of climate change with the histories of "genocide, colonialism, enslavement, and other forms of state and institutional violence that Black communities have endured." The reference to rebellion as "a vision and practice of *overthrowing* the system and liberating people" perfectly fits with the climate insurgency at the very core of our collective reflection.

Roberta Biasillo is assistant professor of contemporary political history at Utrecht University and visiting Max Weber Fellow at the European University Institute in Florence. She has coauthored *Mussolini's Nature: An Environmental History of Fascism* (forthcoming).

Ashley Dawson is professor of English at the Graduate Center and at the College of Staten Island, City University of New York. He currently works in the fields of environmental humanities and postcolonial eco-criticism. He is the author of three recent books relating to these fields: *People's Power* (2020), *Extreme Cities* (2017), and *Extinction* (2016).

Ethemcan Turhan is an assistant professor of environmental planning at the University of Groningen. He coedited *Transforming Socio-Natures in Turkey: Landscapes, State, and Environmental Movements* (2019).

Marco Armiero is research director at the Institute for Studies on the Mediterranean, CNR (Italian National Research Council) and director of the Environmental Humanities Laboratory, KTH–Royal Institute of Technology, Sweden. He is the author of *Wasteocene: Stories from the Global Dump* (2021).

Notes

The Occupy Climate Change! collective would like to acknowledge FORMAS (Swedish Research Council for Sustainable Development) for the support it provided the collective under the National Research Programme on Climate (Contract: 2017-01962_3).

1. *Quint*, "Shaheen Bagh's Bilkis Dadi Detained."
2. Mashal, Schmall, and Kumar, "As Angry Farmers Take to New Delhi's Streets."
3. While we employ the term "climate emergency" here, we also acknowledge that such framings are not without problems; see Nissen and Cretney, "Retrofitting an Emergency Approach to the Climate Crisis."
4. Imran, "Climate Change in the Indian Farmers' Protest."
5. Ahmad, "Climate Crisis Is Foundation of Indian Farmers' Protests."
6. Guha and Martínez Alier, *Varieties of Environmentalism*, 12.
7. Guha and Martínez Alier, *Varieties of Environmentalism*, 13.
8. Hot City Collective, "Hot City."
9. Williams, *Combined and Uneven Apocalypse*.

10. Malm, "Tahrir Submerged?"

11. The literature tracking this opposition is voluminous, including work as diverse as Williams, *The Country and the City*, and Kaika, *City of Flows*.

12. Moore, *Anthropocene or Capitalocene?*

13. United Nations, "Sustainable Development Goal 11."

14. For a detailed discussion of experiences from the Global South and agroecology's role in eco-socialist transformations, see also Ajl, *A People's Green New Deal*.

15. Estes, "A Red Deal."

16. Dawson, *Extreme Cities*.

17. Rice et al., "Against Climate Apartheid."

18. Klein, "Let Them Drown."

19. See, for example, Prewitt et al., "The Politics of Fear," and Gruenewald, "19 Years after 9/11."

20. Millhiser, "Trump's Authoritarian 'Anarchist Jurisdictions' Memo."

21. Manavis, "Eco-fascism."

22. For a study on the political uses of fear on climate change in ordinary spaces, see Bettini, Beuret, and Turhan, "On the Frontlines of Fear."

23. Fearless Cities.

24. Russell, "Beyond the Local Trap."

25. Calbó and Hundal, "Sanctuary and Refuge Cities"; Vaughan and Griffith, "Map."

26. On the new municipalism see Barcelona En Comú, *Fearless Cities*, and Agustin, "New Municipalism as Space for Solidarity."

27. "C40 is a network of the world's megacities committed to addressing climate change. C40 supports cities to collaborate effectively, share knowledge and drive meaningful, measurable and sustainable action on climate change" (C40 CITIES, "About").

28. Gabrys, "Programming Environments," 30–31.

29. Howson, "Climate Crises and Crypto-Colonialism."

30. Hartley, "Smart Cities and the Plight of Cultural Authenticity."

31. Franchise colonialism is generally defined in opposition to settler colonialism. See Yuval-Davis, *Unsettling Settler Societies*, and Shohat, "Notes on the 'Post-Colonial.'"

32. M4BL, "Vision for Black Lives."

33. M4BL, "The Dual Crisis of Climate Change and Covid-19."

34. See, for instance, Cohen, "Confronting the Urban Climate Emergency," and Goh, "Urbanising Climate Justice."

35. Guha and Martínez Alier, *Varieties of Environmentalism*, 13–14.

36. On the experience of self-managed factories see Atzeni and Ghigliani, "Labour Process and Decision-Making" and Azzellini, "Labour as a Commons."

37. Extremely inspirational is the experience of the Sem Terra movement in Brazil; see Hoffmann and Fox, "Food Sovereignty" and Lundström, *The Making of Resistance*.

38. García-Lamarca's "From Occupying Plazas to Recuperating Housing" has explicitly connected the occupation of public spaces in the city with the occupations of buildings to foster affordable housing. On the prefigurative politics coming from squatted houses see Karaliotas and Kapsali, "Equals in Solidarity."

39. An example is the case of the Zone à Défendre (ZAD), an area that was occupied by activists to prevent the construction of a gigantic airport near Nantes, in France. See Bulle, "A Zone to Defend."

40. On this see Mudo, "Introduction: Italians Do It Better?" and Gray, "Beyond the Right to the City."

41. Campbell, "A Critique of the Occupy Movement"; Kilibarda, "Lessons from #Occupy in Canada"; Farrow, "Occupy Wall Street's Race Problem"; Kauanui, "A Structure, Not an Event."

42. Barker, "Manna-Hata."

43. Dawson, *Extreme Cities*, chap. 6.

44. M4BL, "Invest-Divest."

45. Kinniburgh, "Climate Politic after the Yellow Vests."

46. To know more about this accident see Gayle and Quinn, "Extinction Rebellion Rush-Hour Protest." On Twitter, XR wrote, "We are engineers. We are lawyers. We are doctors. We are everyone" (XR UK, "XR Professionals"). XR's idea of who is everyone is quite telling. A radical but also constructive critique of the XR is in Wretched of the Earth, "An Open Letter."

47. In October 2019 an Extinction Rebellion activist sent flowers and a card to the Brixton police precinct where he had been detained to thank the police for their professionalism and courtesy. Precisely at Brixton three Black men had died in custody. The social media discussion following this accident showed even more dramatically XR activists' inability to acknowledge white privilege. See Blowe, "It Is Not Just a Bunch of Flowers."

48. Castán Broto and Robin, "Climate Urbanism," 4.

49. Turhan and Armiero, "Of (Not) Being Neighbors."

References

Ahmad, Omair. "Climate Crisis Is Foundation of Indian Farmers' Protests." *Eco-Business*, January 25, 2021. https://www.eco-business.com/news/climate-crisis-is-foundation-of-indian-farmers-protests/.

Agustin, Oscar Garcia. "New Municipalism as Space for Solidarity." *Soundings*, no. 74 (2020): 54–67.

Ajl, Max. *A People's Green New Deal*. London: Pluto, 2021.

Atzeni, Maurizio, and Pablo Ghigliani. "Labour Process and Decision-Making in Factories under Workers' Self-Management: Empirical Evidence from Argentina." *Work, Employment and Society* 21, no. 4 (2007): 67–86.

Azzellini, Dario. "Labour as a Commons: The Example of Worker-Recuperated Companies." *Critical Sociology* 44, nos. 4–5 (2018): 763–76.

Barcelona En Comú, eds. *Fearless Cities. A Guide to the Global Municipalist Movement*. Oxford: New Internationalist Publications, 2019.

Barker, Joanne. "Manna-Hata." *Tequila Sovereign* (blog), October 10, 2011. https://tequilasovereign.wordpress.com/2011/10/10/manna-hata/.

Bettini, Giovanni, Nicholas Beuret, and Ethemcan Turhan. "On the Frontlines of Fear: Migration and Climate Change in the Local Context of Sardinia, Italy." *ACME: An International Journal for Critical Geographies* 20, no.3 (2021): 322–40.

Blowe, Kevin. "It Is Not Just a Bunch of Flowers." *Medium*, October 16, 2019. https://medium.com/@copwatcher_uk/it-is-not-just-a-bunch-of-flowers-bc5078b899e4.

Bulle, Sylvaine, "A Zone to Defend: The Utopian Territorial Experiment of Notre Dame des Landes." In *Everyday Resistance: French Activism in the 21st Century*, edited by Bruno Frère and Marc Jacquemain, 205–28. Cham, Switzerland: Palgrave Macmillan, 2020.

Calbó, Ignasi, and Sunny Hundal. "Sanctuary and Refuge Cities." OpenDemocracy,

June 14, 2017. https://www.opendemocracy.net/en/can-europe-make-it/sanctuary-and-refuge-cities/.

Calder Williams, Evan. *Combined and Uneven Apocalypse: Luciferian Marxism*. Winchester: Zer0 Books, 2011.

Campbell, Emahunn Raheem Ali. "A Critique of the Occupy Movement from a Black Occupier." *Black Scholar* 41, no. 4 (2011): 42–51.

Castán Broto, Vanesa, and Enora Robin. "Climate Urbanism as Critical Urban Theory." *Urban Geography*. Published ahead of print, November 17, 2020. https://doi.org/10.1080/02723638.2020.1850617.

Cohen, Daniel Aldana. "Confronting the Urban Climate Emergency: Critical Urban Studies in the Age of a Green New Deal." *City* 24, nos. 1–2 (2020): 52–64.

C40 CITIES. "About." https://www.c40.org/about (accessed March 1, 2021).

Dawson, Ashley. *Extreme Cities: The Peril and Promise of Urban Life in the Age of Climate Change*. London: Verso, 2017.

Estes, Nick. "A Red Deal." *Jacobin*, August 6, 2019. https://www.jacobinmag.com/2019/08/red-deal-green-new-deal-ecosocialism-decolonization-indigenous-resistance-environment.

Farrow, Kenyon. "Occupy Wall Street's Race Problem. White Protesters Need to Rethink Their Rhetoric." *American Prospect*, October 24, 2011. https://prospect.org/civil-rights/occupy-wall-street-s-race-problem/.

Fearless Cities. https://www.fearlesscities.com (accessed March 1, 2021).

Gabrys, Jennifer. "Programming Environments: Environmentality and Citizen Sensing in the Smart City." *Environment and Planning D: Society and Space* 32, no. 1 (2014): 30–48.

García-Lamarca, Melissa. "From Occupying Plazas to Recuperating Housing: Insurgent Practices in Spain." *International Journal of Urban and Regional Research* 41, no. 1 (2017): 37–53.

Gayle, Damien, and Ben Quinn. "Extinction Rebellion Rush-Hour Protest Sparks Clash on London Underground." *Guardian*, October 17, 2019. https://www.theguardian.com/environment/2019/oct/17/extinction-rebellion-activists-london-underground.

Goh, Kian. "Urbanising Climate Justice: Constructing Scales and Politicising Difference." *Cambridge Journal of Regions, Economy, and Society* 13, no. 3 (2020): 559–74.

Gray, Neil. "Beyond the Right to the City: Territorial Autogestion and the Take over the City Movement in 1970s Italy." *Antipode* 50, no. 2 (2018): 319–39.

Gruenewald, Jeff, Joshua D. Freilich, Steven Chermak, and William Parkin. "19 Years After 9/11, Americans Continue to Fear Foreign Extremists and Underplay the Danger of Domestic Terrorism." *Conversation*, September 10, 2020. https://theconversation.com/19-years-after-9-11-americans-continue-to-fear-foreign-extremists-and-underplay-the-dangers-of-domestic-terrorism-145914.

Guha, Ramachandra, and Juan Martínez Alier. *Varieties of Environmentalism: Essays North and South*. 1998; repr., London: Earthscan, 2006.

Hartley, Kris. "Smart Cities and the Plight of Cultural Authenticity." *Global Urbanist*, March 24, 2015. https://globalurbanist.com/2015/03/24/smart-cities-cultural-authenticity.

Hoffmann, Ester, and Michael Fox. "Food Sovereignty, One Occupation at a Time." *NACLA Report on the Americas* 49, no. 4 (2017): 451–56.

Hot City Collective. "Hot City Compound Crisis and Popular Struggle in NYC." Verso Blog, August 3, 2020. https://www.versobooks.com/blogs/4811-hot-city-compound-crisis-and-popular-struggle-in-nyc.

Howson, Peter. "Climate Crises and Crypto-Colonialism: Conjuring Value on the

Blockchain Frontiers of the Global South." *Frontiers in Blockchain*, May 13, 2020. https://doi.org/10.3389/fbloc.2020.00022.

Imran, Zafar. "Climate Change in the Indian Farmers' Protest." *Le Monde diplomatique*, February 1, 2021. https://mondediplo.com/outsidein/climate-indian-farmers.

Kaika, Maria. *City of Flows: Modernity, Nature, and the City*. New York: Routledge, 2012.

Karaliotas, Lazaros, and Matina Kapsali. "Equals in Solidarity: Orfanotrofio's Housing Squat as a Site for Political Subjectification across Differences amid the 'Greek Crisis.'" *Antipode* 53, no. 2 (2020): 399–421. https://doi.org/10.1111/anti.12653.

Kauanui, J. Kehaulani. "'A Structure, Not an Event': Settler Colonialism and Enduring Indigeneity." *Lateral. Journal of the Cultural Studies Association* 5, no. 1 (2016). https://csalateral.org/issue/5-1/forum-alt-humanities-settler-colonialism-enduring-indigeneity-kauanui/.

Kilibarda, Konstantin. "Lessons from #Occupy in Canada: Contesting Space, Settler Consciousness, and Erasures within the 99 Percent." *Journal of Critical Globalisation Studies* 5 (2012): 24–41.

Kinniburgh, Colin. "Climate Politics after the Yellow Vest." *Dissent*, Spring 2019. https://www.dissentmagazine.org/article/the-yellow-vests-uncertain-future.

Klein, Naomi. "Let Them Drown." *London Review of Books*, June 1, 2016. https://www.lrb.co.uk/the-paper/v38/n11/naomi-klein/let-them-drown.

Lundström, Markus. *The Making of Resistance: Brazil's Landless Movement and Narrative Enactment*. Cham, Switzerland: Springer, 2017.

Malm, Andreas. "Tahrir Submerged? Five Theses on Revolution in the Era of Climate Change." *Capitalism Nature Socialism* 25, no. 3 (2014): 28–44. https://doi.org/10.1080/10455752.2014.891629.

Manavis, Sarah. "Eco-fascism: The Ideology Marrying Environmentalism and White Supremacy Thriving Online." *New Statesman*, September 21, 2018. https://www.newstatesman.com/science-tech/2018/09/eco-fascism-ideology-marrying-environmentalism-and-white-supremacy.

Mashal, Mujib, Emily Schmall, and Hari Kumar. "As Angry Farmers Take to New Delhi's Streets, Protests Turn Violent." *New York Times*, January 26, 2021. https://www.nytimes.com/2021/01/25/world/asia/india-farmers-protests-delhi.html.

Millhiser, Ian. "Trump's Authoritarian 'Anarchist Jurisdictions' Memo, Explained." *Vox*, September 3, 2020. https://www.vox.com/2020/9/3/21419767/trump-anarchist-jurisdictions-authoritarian-illegal-unconstitutional-supreme-court.

Moore, Jason W. *Anthropocene or Capitalocene? Nature, History, and the Crisis of Capitalism*. San Francisco: PM/Kairos, 2016.

Mudu, Pierpaolo, "Introduction: Italians Do It Better? The Occupation of Spaces for Radical Struggles in Italy. Symposium: The Occupation of Spaces for Radical Struggles in Italy Organizer: Pierpaolo Mudu." *Antipode* 50, no. 2 (2018): 447–55.

M4BL. "The Dual Crisis of Climate Change and Covid-19." https://m4bl.org/events/the-dual-crisis-of-climate-change-and-covid-19/ (accessed March 1, 2021).

M4BL. "Invest-Divest." https://m4bl.org/policy-platforms/invest-divest/ (accessed March 1, 2021).

M4BL. "Vision for Black Lives." https://m4bl.org/policy-platforms/ (accessed March 1, 2021).

Nissen, Sylvia, and Raven Cretney. "Retrofitting an Emergency Approach to the Climate Crisis: A Study of Two Climate Emergency Declarations in Aotearoa New Zealand." *Environment and Planning C: Politics and Space*. Published ahead of print, June 28, 2021. https://doi.org/10.1177/23996544211028901.

Prewitt, Kenneth, Eric Alterman, Andrew Arato, Tom Pyszczynski, Corey Robin, and Jessica Stern. "The Politics of Fear after 9/11." *Social Research* 71, no. 4 (2004): 1129–46.

Quint. "Shaheen Bagh's Bilkis Dadi Detained for Joining Farmers' Protests." December 1, 2020. https://www.thequint.com/news/india/shaheen-baghs-bilkis-dadi-detainedfor-joining-farmers-protests.

Rice, Jennifer L., Joshua Long, and Anthony Levenda. "Against Climate Apartheid: Confronting the Persistent Legacies of Expendability for Climate Justice." *Environment and Planning E: Nature and Space*. Published ahead of print, March 12, 2021. https://doi.org/10.1177/2514848621999286.

Russell, Bertie. "Beyond the Local Trap: New Municipalism and the Rise of the Fearless Cities." *Antipode* 51, no. 3 (2019). https://doi.org/10.1111/anti.12520.

Shohat, Ella. "Notes on the 'Post-Colonial.'" In *The Pre-Occupation of Post-Colonial Studies*, edited by Fawzia Afzhal-Khan and Kalpana Rahita Seshadr, 126–39. Durham, NC: Duke University Press, 2000.

Turhan, Ethemcan, and Marco Armiero. "Of (Not) Being Neighbors: Cities, Citizens, and Climate Change in an Age of Migrations." *Mobilities* 14, no. 3 (2019): 363–74.

United Nations. "Sustainable Development Goal 11: Sustainable Cities and Communities." https://www.un.org/sustainabledevelopment/cities/ (accessed March 1, 2021).

Vaughan, Jessica M., and Bryan Griffith. "Map: Sanctuary Cities, Counties, and States." *Center for Immigration Studies*, March 22, 2021. https://cis.org/Map-Sanctuary-Cities-Counties-and-States.

Williams, Raymond. *The Country and the City*. 1973; repr., London: Vintage, 2016.

Wretched of the Earth. "An Open Letter to Extinction Rebellion." *Red Pepper*, May 3, 2019. https://www.redpepper.org.uk/an-open-letter-to-extinction-rebellion/.

XR UK. "XR Professionals are at Trafalgar Square now in our 1000s. . . ." Twitter, October 17, 2019, 1:55 p.m. https://twitter.com/xrebellionuk/status/1184890614401458176.

Yuval-Davis, Nira. *Unsettling Settler Societies: Articulations of Gender, Race, Ethnicity, and Class*. London: Sage, 1995.

Majority Urban Politics and Lives Worth Living in a Time of Climate Emergencies

AbdouMaliq Simone and Solomon Benjamin

In the hardscrabble cities of the so-called Global South, what constitutes a viable politics of climate mitigation for an urban majority sometimes keeping its head above water and, at other times, creating the semblance of dynamic livelihoods? Despite enduring poverty, there is much about urban life in the South that is not simply reducible to the compounding of hardship and a substantial expansion of material consumption. Over this tremulous in-between hover the substantive signs of deleterious climate change. The interlaced relations among global warming, rising sea levels, floods, and severely polluted and overheated urban atmospheres, as well as the insufficient ways these are often compensated for, present critical challenges to the very capacity to prolong the inhabitation of many urban regions.[1]

Across the world, these challenges become the locus of new modalities of urban activism, where the right to the city is converted into a right to continue living.[2] They also raise important questions as to the contours of life worth living and the extent to which that life is viably framed within the confines of "the human." If much of the politics of climate mitigation is undertaken in terms of preventing the extinction of human life, of constantly invoking the threats of climate change to the very survival of the human species, to what extent does this invocation occlude the ways in which the human has dismissed and foreclosed ways of living for those long deemed ineligible for a fully human life? Of course, the human largely functions as a pragmatically necessary universalism that attempts to provide some workable standard through which diverse persons can

Social Text 150 · Vol. 40, No. 1 · March 2022
DOI 10.1215/01642472-9495089

be accountable to one another beyond the specificities of the context of their inhabitation. The human, as Eugene Thacker alludes, may simply be another way for some other agent of existence to perpetuate itself. But what seems salient here is both how the universalism, defined largely through white and colonial privilege, is both embraced and refused by those who are the primary recipients of judgments regarding the importance of sustaining human life—that is, the poor.[3]

So, how does this invocation, now often expressed through transnational constructions of human rights, relate to the murky everyday life politics of living in urban districts whose heterogeneities, conflicts, and complicities tend to play out in terms that could be construed as indifferent to a specifically human development or sustenance? How does configuring everyday spaces of operation entail an often down-and-dirty series of accommodations, brokering, wheeling and dealing, and tradeoffs aimed not so much at protecting the integrity of particular human identities as at elaborating a worthiness of life based on the capacity to "roll with the punches," to make something out of whatever presents itself, no matter how toxic it might appear? As the disciplinary regimens of urban modernity imposed in postcolonial cities often emphasized the need for a human development predicated on relinquishing what was perceived as the unruliness and incivility of the masses, how do poor districts today attempt to recuperate a sense of the collective forms of living once valued as capable of creating a "real urban life"?

Additionally, what role does land play in these "roll with the punches" everyday lives? As land is embedded in claims and occupancy, and as land administration spurs and politicizes a politics of living whose premise informs and gains advantage in the resultant composition of operational territories, who can do what with whom and when?

These questions and dilemmas will be addressed through considering some of the characteristics and practices of pro-poor politics in the district of Penjaringan in northern Jakarta and the particular ways in which low-income residents occupy land in Bangalore. These small case studies illustrate the complexities entailed in attempting to manage environmental change amid seemingly overwhelming obstacles. The essay then goes on to consider possible reformulations of a life worth living beyond the confines of the normative idioms of the human and what implications this might have for emerging urban politics in the future.

Part One

Walling Off the Sea (and Everything Else)

In the northern Jakarta districts of Muara Baru and Penjaringan, the ground is literally sinking under the feet of the poor. But faced with years and years of evictions and threats of eviction, the poor are holding their ground, even as subsidence threatens to undermine substantial political gains.[4] During the heated gubernatorial campaign of 2017, the victorious candidate, Anies Baswedan, managed to capture a critical segment of the "poor vote" by entering into a political compact with Jaringan Rakyat Miskin Kota and the Urban Poor Consortium, two leading coalitions of organizations of the urban poor, that committed the new administration to a policy of no evictions and in situ redevelopment through participatory planning processes. During the previous administration of Basuki Purnama (Ahok) there had been 306 forced evictions, most conducted under the auspices of environmental security.

It is cruelly ironic that poor residents had been targeted in this regard as northern Jakarta has witnessed the prolific rollout of high-end vertical residential developments on former swamp land that extract deep groundwater and harden the riverine deltas that are outlets for the scores of rivers that empty into the Jakarta Bay. These actions not only substantially deprive poor residents of their water supply but also largely contribute to subsidence levels now averaging fifteen centimeters per annum.[5]

The situation has been exacerbated by the fitful starts and stops of major redevelopment projects intended for the Jakarta Bay. In a region plagued with flooding, rising sea levels, subsidence, rampant overdevelopment, reticulations to bulk water supply limited to only about 30 percent of the population, and clogged rivers, it is widely assumed that North Jakarta could literally be wiped out within five decades. While Suharto's New Order regime had toyed with the idea of land reclamation and the building of artificial islands in the Jakarta Bay as part of turning Jakarta into a world-class city as far back as 1995, the reclamation project that has been in the works for the past decade intends to save the city from flooding and is a critical component of Indonesia's climate mitigation strategy.[6]

In what was costed as the world's most expensive infrastructure project, a sea wall is to enclose the bay, turning it into a freshwater reservoir, equipped with an elaborate pumping system to remove effluvium. The enormous costs of the project are to be partially funded through the development of sub-cities on a series of nine artificially created islands, conceded to the city's largest real estate developers. These developers have also been significant funders of the dredging and widening projects of the area's rivers and reservoirs, affecting the poor particularly, as their settlements are frequently situated on riverbanks. While this project is neither proceeding nor completely abandoned, a temporality of constant and uncertain adjustments is aimed simply at buying time.

The seawall and related projects, under the auspices of the National Capital Integrated Coastal Development, is largely driven by a consortium of Dutch consulting firms in collaboration with the Ministry of Public Works and Ministry of Development Planning in a partnership that reiterates many aspects of colonial relations. Because the city of Jakarta retains certain jurisdictional powers over this area in terms of land-use regulations, a range of conflicts among institutional actors and the heavy reliance upon the expertise and capital of the city's major private development conglomerates has led to several incidents of illegal construction, such as the almost overnight construction of nearly three hundred shophouses on Island G in 2017.[7]

North Jakarta has been subject to scores of plans of rectification.[8] For example, Dutch funding drove the formation of the Human Cities Coalition for Jakarta and Manila in 2016, which deployed the language of pro-poor sustainable development to roll out so-called hybrid housing for low-income residents of Penjaringan—a way of amalgamating the designs of self-constructed housing within a small-scale vertical platform.[9] But the project prompted its own debilitating speculative pressures on informal land markets, and many poor residents for which these developments were intended simply withdrew any kind of cooperation, sensing that these were simply a prelude to other transformations that would undermine their economic position and long-term security. There is a general absence of consideration of the ways in which different "kinds of water" are connected to each other, a lack of institutional sense of what water does, and how it ramifies across various social and infrastructure relations.[10]

The problem with many projects of "rectification" or "normalization" is that they are instantiated in an urban fabric of intense heterogeneity in the compositions of both the built and social environment.[11] Even for poor residents, their living situations and physical platforms vary widely, from self-built and makeshift wood constructions clinging to the sides of riverbanks, to densely crowded blocks of three- to four-story "tenements," to makeshift dormitory structures within abandoned warehouses and factories.[12] Penjaringan is a jumble of superblocks, high-end suburban estates, old stocks of social housing largely expropriated by lower-middle-class households even as all kinds of improvised arrangements are required to service flats with water and power, large swathes of well-built, stylish multistory residences, rows upon rows of combined residential-commercial shophouses, as well as the modest, basically well-constructed two-story edifices that house working-class or working-poor families.

Combined with thousands of shops, small factories, workshops, markets, shopping malls, interspersed with all kinds of *terroir vague,* almost every kind of economic activity or social categorization imaginable is packed into a highly dense area. In order for all these multiplicities to coexist under conditions where opportunities for consensual deliberation are minimal or inconceivable, they must all work their way around one another, and cultivate all kinds of physical and cultural buffer zones.[13] Political organization is distributed across a variety of authority figures and forms of mobilization that often don't mirror our familiar forms of mobilization and activism. Still an intricate choreography of popular local forces is at work, and it is often difficult to know for sure what it accomplishes beyond enabling many of the urban poor to quietly remain in place.[14] This then becomes a critical conundrum in assessing viable forms of political mobilization for the urban poor. For the sheer intricacy of the intersections among daily transactions, among actors with different resources and capacities, affords the poor a continuous series of "strategic openings" to improve livelihood and social conditions, but without the necessary reversing of structural conditions of their relative impoverishment. Yet, too many political interventions tend to abstract the poor from these heterogenous arrangements, diminishing the horizons of political possibility in the very exigency of making their presence and needs more visible and specific.

Commonality Always to Be Reinvented

Much has been made in cities across the world about middle-class and poor coalitions centered on environmental protection. In Penjaringan, despite many women and youth organizations that cooperate with one another to advocate for rights and better livelihoods, the primary concern of the "majority" is at all costs not to be as poor as those "on the other side of the streets." This tends to militate against "cross-class," "cross-micro-territory" affiliation because one just cannot afford to be seen as having something to do with that which is even infinitesimally "lower" than his or her situation.[15]

At the same time, this does not mean that collaborative activity is not happening all the time.[16] Perhaps with the exception of upper-middle-class and wealthy residents ensconced in the upscale Pluitt subdistrict of Penjaringan, the district is replete with complicities of all kinds, from accessing work, services, favors, and support. The poor and middle class are thoroughly imbricated with each other's lives in the way they avail to each other their relative comparative advantages—in terms of the knowledge of the street, about where certain resources can be obtained, about informal work, about how to fix various problems. For the consolidation of the position of each, what each has to do in terms of putting food on the table, of knowing what is going on across the district, provides certain resources for each that the other does not have ready access to.[17]

These discrepancies, far from constituting only the markers of separation, become modalities for implicit exchange.[18] This situation is aided through retaining some fundamental ambiguities about to whom things really belong, whether this is in terms of specific tracts of land or public space. Efforts to render these ambiguities more transparent, to straighten them out, thus undermine an important "platform" of interconnectedness among disparate groups and interests. While these implicit exchanges among diverse residents are an integral aspect of sustaining the residents' presence with one another in ways that temper conflict, their motivation largely remains on the level of protecting self-interest, not the well-being of a "community" as a whole. Here, the poor have an important degree of agency, but not as humans whose lives pale in sufficiency compared to others with whom they share a common district, but rather as a complex field of antagonisms and diverse interests in which the presence and action of the poor constitutes an important element in the working out of affordances, labor, strategic knowledge, and resources. While there are no guarantees against eviction or a massive remaking of the existing landscape, the intricacies of Penjaringan's interweaving of social and built environments has in most respects proven to be an effective shield to a wholesale uprooting of the life of most of the working poor in this district. Deeply embedded in multiple circuits of everyday economic exchange, it is this very embedding, rather than the amplification of the specific needs of the poor themselves, that offers limited guarantees.

Despite long-term efforts to build consortia of the urban poor, to think in terms of such a whole, a more collective "we" raises issues of belonging and entitlement that the political culture of Jakarta has contributed little to working through.[19] This is reflected in popular discussions about where residents are from, where they "belong to." It is rare to hear people say they are from Jakarta even if they have been born and lived there all their lives, for they will inevitably cite the place of origin of a parent or more usually grandparent, even when their exposure to that place may be limited to at most an annual visit during the Eid al Fitr holiday. While all speak an intensely Jakarta dialect of the national language, have no interest in living anywhere else, and are likely to host a slew of distant relatives migrating to the city, they express few ideas about what it would mean to "belong" to Jakarta. While the city is a vast reservoir of diverse ways of being, for which they usually express their appreciation, the horizons of what is important for them usually remain intensely local, even parochial.

A long-term practice, especially among poor and working-class residents, is a begrudging disregard for local transgressions, and at least a feigned indifference to the practices of neighbors that might diverge from their own, as well as a willingness to let them go. In other words, the capacities of people to reside together in close proximity without clear ideas of a sense of mutual belonging have depended upon a willingness not to see the behavior of others as having something to do with one's own capacities to carve out a viable life, even as those differences are often converted into the basis of some kind of complementary exchange. So, while the environmental precarities of Penjaringan are enormous and the long-term prospects for continued residency in the area in question, even with plans to move the capital from Jakarta, it is difficult to conceive of a politics that might enjoin its diverse residents into an overt, as opposed to a tacit, form of sustained collective action.

That said, limited campaigns initiated on the part of the basic local governmental units of the district, often as competitions among them, to "green" neighborhoods through the introduction of plants, the sorting of waste, and the appropriate disposal of litter are often successful, as they operate under the guise of extended-family projects. They become supplements to the activities already engaged in on a daily basis, which neighbors always witness and thus are viewed as a kind of immediacy, not challenging the basic presumptions that neighbors have about one another.[20] For crisis is not viewed as something brewing over time, as something that can be anticipated or prepared for, and thus the fact that the ground is literally sinking under the feet of Penjaringan residents, even if overtly registered in recognition of increased flooding, of the appearance of cracks in the wall and on pavements, is perceived as something that can be used against them by more powerful actors. There are parts of Penjaringan along the bay where poor families are constantly being flooded out of their makeshift, box-carton constructions, and yet they constantly return to the "scene of the crime," in part because their choices are limited. But they also return, in part, because this is simply "where they are," as one resident expressed it; they have become part of this landscape in a near symbiotic relationship; they have been extended into it as it extends itself to them.

This, in turn, raises fundamental questions about who and what is to be sustained in any environmental, climate change politics. In the long banal reign of Suharto's New Order, residents were to accommodate themselves to a fundamental disregard of their well-being as the state promised to take care of them. The educational system had no interest in promoting inquiry, in students thinking for themselves. In a generalized formatting of the population, everyone was expected to do the right thing, to focus everything on generating normative appearances for general consumption. There was little pretense for anything that might promote a human flourishing. As such, what does the sustainability of human life mean in such a context in which, for decades, such life was basically reduced to a kind of collective biomass subjected to a vagary of standardized rituals.[21] At the same time, Penjaringan, despite its precarities, is a testament to the skilled practice of a kind of collective politics that produces an interweaving of different ways of life but lacks an incisive vernacular capable of representing this capacity to itself. This is why local activists have focused recently on devising nightly roving public dinners across the district, providing an elemental public image of that interweaving.

A Human Project? Brokering a Life Worth Living

The *kampung* has been stereotyped as the village in the city, the reproduction of village mores as the means of social reproduction in an urban context that otherwise did not provide sufficient institutional resources for such social reproduction. It was seen by the elite and middle class as the concrete evidence that the "majority" of Jakarta's residents remained ineligible for "real" urban citizenship. In this imagination, the survival of the *kampung* has been predicated on the absence of any widespread collective sensibility of the characteristics and responsibilities of an "urban human." In other words, these were not residents capable of "possessing themselves," of demonstrating propriety through regarding themselves as a property to be developed and maintained.

The poor have long been vilified as something that must be kept apart, relegated to the *kampung,* something not fully human, in the terms that an urban elite and burgeoning middle class came to understand that designation. As Neferti Tadiar points out, the predominant condition of our age is the war between those who are attempting to remain human at all costs and those who will probably never attain such a status, who simply fight to retain a semblance of a life worth living—something that is rarely construed as a "human project."[22] It is this notion of a life worth living that is often occluded from the framing of climate change politics. For Tadiar, it is not that the poor necessarily refuse the sense of being human, but rather that the aspiration for human dignity and human necessity has often been mobilized to denigrate the very skills and capacities of the poor, that the very ways in which they attempt to do more than survive their "bare" human life is dismissed as insufficiently human. In other words, the poor come to refuse the "refusal," and thus implicitly expand the terrain of human—that is, its ability to extend, even disappear across multiple relations with the more than human landscape or built environment. What might be viewed as toxic or self-destructive is revalorized as a work in progress.

In Penjaringan, the oscillations of class conflict and complementarity, the compaction of heterogeneities into a constantly unsettled interchange of divergent interests and ways of doing things, and the rampant disregard for the well-being of long-term residents manifested by speculative real estate capital and the self-aggrandizing brutalities of an elite all are undertaken with few considerations of human rights or any kind of specifically *human* project. In part stifled by the style of new order governmentality that sequestered a political imagination in terms of increased material consumption and in part simply a by-product of residing in a city with few official affordances, there is a limited horizon of aspiration for most Jakarta residents I have come to know over the years. For the orientation of everyday life is basically geared toward negotiating obstacles—all the things that get in the way of getting the basic things done, of getting from here to there in interminable traffic, where the present needs to be constantly repaired, leaving little for planning for a future.[23] At the same time, aspirations, especially to attain middle-class status, are increasingly perceived as a trap, a lure into indebtedness, and a distraction from efforts to keep all options open.

Part Two

Contesting Dangerous Spaces

Climate change does not simply concern rising sea levels, and the subsequent floods and landslides. Rather an apparatus of governance is produced with planning documents, newspaper articles, and technical studies employing Geographic Information Systems (GIS) and GPS data to constitute the details of transformation, of what is potentially dangerous, along with international seminars, workshops, and classrooms where courses on climate change and sustainability are conducted in techno-rational and evangelical ways. The effort to render territory safe and secure thus contrasts with the more liminal aspects of territory formation that have been crucial platforms for the livelihood of the majority.

For example, many urban spaces embody a "spiritual logic" that is also brought to bear on land claims and development. Stories concerning Indian urban development are replete with how eight lane highways came to a mysterious halt or how the collapse of new buildings were not sufficiently "insured spiritually."[24] "Sacredness" is constructed via the embedded experience of the family rather than the usual assumptions of its being "religious." A particular place may embody an intense symbolic value because it is the site where a mother may have revealed the stories of a long genealogy to her offspring, or where a sudden death might have occurred. When sacred deities are also legal subjects, their status shapes claims to land from logic that spans across both space and time and yet is presented in a highly materialized form every day.

In South Bangalore's trading spaces, wealthy resident welfare associations (RWAs) have mobilized various legal instruments to undermine the residency of mostly a Thigla caste of flower cultivators and sellers who settled in the Maystripalya district during the 1930s. These RWAs of the Bangalore Development Authority were formed in the 1980s, and have long labeled groups such as the Thigla as encroachers that form slums that threaten law and order. Low-income and working-class residents that have for nearly a century demonstrated a measure of economic vitality and viability are increasing regarded as constituting what the RWAs label a "wild scape," making territory "dangerous."

The efforts of these RWAs have erased three of five Thigla sacred sites, which are headed by female deities as "seven sisters" that form a territorial configuration organized centrally around a *kere* wetland. This is now bifurcated into a landscaped lake, whose underground includes three stormwater drains emptying the effluvium of the posh 3rd Block into a chain-link-fence-circumscribed marsh impeding grazing once accessed by the Maystripalya settlers. Local Thigla leaders associated with one of the many arms of the Dalil Sangarsh Samiti (DSS), recounting this act as a "holocaust," mobilized their alliances with both elderly corporators and administrators within the Bangalore City Corporation's elected councils to lobby the province-level revenue and irrigation administration, under which such "minor" wetlands fall, so as to restitute the traditional uses of this land.

These contests between a resident welfare association—largely made up of US returnees, NRIs (Non-Resident Indians) attempting to recreate their Florida and Californian dreams—and differently rooted Thigla prompt the latter to initiate "upgrades" of their own. Here a cluster of concrete-block shack housing draws on "scheme" funds to cater to what the Thigla leadership sees as ways to relieve congestion within Maystripalya. Concrete paved lanes are densified by lines of plastic cans that store a day's supply of water from single pipelines. The pipelines, the concrete paving, and now the newly built shacks reflect Thigla politicization of administrative spaces. The shacks are of course useful to address the overcrowding within the core settlement. But these are also intended to safeguard the land as a territorial extension against the incursions sought by the 3rd Block RWA to transform wetlands into "lakes" proposed as ecozones. An eight-foot-high wall surrounds the concrete shacks.

A key symbol of this reclamation of territory is a distinctive memorial to the renowned Dalit intellectual and civil rights campaigner B. R. Ambedkar that lies beyond the concrete-wall shacks and their containing wall. Its location defines the northern part of a muddy walking path around the marshy part of the remaining wetland. Technically, the memorial with its powerful caste connotation is a warning to the mostly upper-caste RWAs and their morning walkers using this path for jogging. This is on land still administered by the minor irrigation department that largely relishes maintaining the last shreds of legitimacy to do so. While the claims made by the elite RWAs rely on cartography (maps and plans, surveys, and the landscape plan to reclaim this space as an "ecological commons"), those by the Mastripalya residents are ensconced in the world of the Ambedkar aura and artifacts, of copper plates given by the British colonial administration to the settlement founder in return to his services, one of which, ironically, was to survey the outer precincts of Bangalore.

The Ambedkar memorial, shaded as a sitting space for lovers and a meeting point for local youths and leaders, can also be read as a politics of using religious shrines to safeguard against demolition. It is not clear whether this will be viable. Further north, Viveknagar, located as Koramangala edges central Bangalore's Victoria Layout, has witnessed repeated demolitions and evictions. Originally a marshland, and in some parts still so, a *kere* was filled up to form the National Games Village but also where poorer settlers were moved as the third resettlement of those affected by a series of evictions in the urban core. At the heart of Viveknagar is a new mall and office complex, encircled by low-income homes, shops, evangelical churches, another, larger Ambedkar shrine, and several temples housing "local" deities with their protective Naga Cobra hooded icons. The priest of the cobra shrine told us that these sacred places form a "fence" against demolitions.

Across northern Bangalore, protective deities encircle the major *keres* that now lie converted into lakes, which become the basis of experimenting with different forms of management and serve as proof that municipal authorities are attempting innovative ways of governing but without these experiments being used to inform a broader set of sites and issues.[25] The *keres* that surround Bangalore, far from being pristine, instead form part of complicated set of eco-agro practices that are spatialized in temporal ways. This includes a shifting territory around the core wetland that as a fluid territory is designated for multiple and sequential uses: the drying of agricultural products, the collection of silt via silk dredging, and fishing.[26] They are hardly "informal" but have long been embedded into village council administrative procedures reflected in titles and also moved around to other parts of the administrative apparatus, built over and merged, and at times replaced by other procedures.

Although individual village and municipal councils have long shaped and internalized these territorial practices, the interlinked wetlands get reterritorialized in the expanding interactions among multiple developers and village and municipal councils. They take the form of settlement extensions, special economic zones, plots for public housing, as well as high-end "gated communities," crisscrossed and often intersected with expressways or smaller roads. These shape intense contestations between different administrative-politico sectors: master planning and mega-project development confront and overlap with extensions of infrastructure and services to expanding low- and mixed-income settlements carved within village boundaries, which themselves oscillate according to different administrative jurisdictions. The valorization of real estate potentials, mostly by RWAs, municipal authorities, developers, and even sometimes by low-income communities, is a critical driver in remaking territory. This is not just in terms of inward flows of large capital investment both domestic and foreign.[27] There are also hundreds of small land developers that partner with village clans to play the land market by developing cheap rental walkup blocks catering to the lower- and middle-end IT workers seeking employment in Bangalore's second "IT corridor."

These real estate surpluses fuel an important measure of political autonomy to village councils and residents, who also subdivide and lease out larger agriculture parcels to contest, via their long-honed connections within administrative and judicial circles, actions by the planning authority that seek to expropriate land to form special economic zones and public, mass housing. However, it would be a mistake to view real estate as the only driver and logic. When the Maystripalya Thiglas used the term "holocaust" in referring to the RWA appropriation of their wetland, they referred to the threat posed to territorial practices that have given meaning to their lives through the rituals connected to specific *kula* (clan) and *gram* (patron) *devathas* (deities) that bring them together. The territorial usurpation is reinforced by the way in which English-speaking Koramangala 3rd Block RWAs derogate the Thigla language as the sounds of dogs barking as a means of legitimating their reluctance to engage in negotiations around land use with them.

While low-income residents, especially those emplaced in particular areas for many decades, are attentive to the changing ground that they inhabit—that is, the vicissitudes of human-induced climate change—they are also wary of the modalities of rectification that put the onus of radical life changes largely on them. For those who have long depended on straddling the murky boundaries between rationality and irrationality, transparent planning/accountability and often murky backroom deals deep within administrative and political bureaus, and between the apparent clarity/definitiveness of law and regulation and the ambiguous productions of multiple and countervailing legitimacies, policy directives based on climate exigencies may have little traction within majority communities. They may be construed as more threat than promise.

Faced with the prospect of emergencies that seem to leave little room for negotiation, more extensive climate activisms will emerge from the grassroots of cities such as Bangalore only by amplifying the negotiability of the shape of needed transformation. For negotiability is at the heart of the capacity for improvisation, and improvisation upon existent, usually discordant urban spatial arrangements of diverging logics, finance, class interests, speculations, and land uses is the only viable process to work toward urban climate justice. Major infrastructural investments and reorganization energy systems are of course necessary. But in cities like Jakarta and Bangalore, the production of space has always entailed a messy politics of negotiation, and it is the rollout of advanced technical and computation systems, often in the name of responding to climate emergency, that threatens to close down negotiability.

Conclusion

A critical question in this time of climate emergency is the extent to which it is important to retain a sense of humanity at all costs. For if the possibility to be human, as a self-reflecting, autonomous subject demonstrating a capacity for free will, is that which is to be defended, then it cannot separate itself from the possibilities of war. If the city has been the locus for the constitution of the very notion of "the human," then it required a population that was excluded from the possibility of being human. The capacity to reflect, deliberate, and invent, all tropes closely associated with the notion of a human that possesses no fundamental nature or character beyond that which can be invented, required a substrate of persons that indeed did have a definitive nature, as beasts of burden, as sheer labor.

Historically, as that sheer labor "comes into its own," struggles against the presumptions of its own absence of eligibility, everyday "wars" of position become inevitable, as humanity is experienced as privilege in *relation* to something else, and has to be demonstrated in relation to a nonhuman capable of acknowledging that privilege. Of course this is a war in which the dispossessed are always at a disadvantage, and while the aspirations to be included in the realm of the human on their own terms provide a platform of legitimation for such struggles, as well as a potentially unifying objective, the dilemma has been how to circumvent the need to be human, in its imposed form, in order to have a life worth living. This life worth living has, for an urban majority in the Global South, been attained not so much from a sense of human flourishing but from a mangle of negotiations and deals that have managed to engender a mangle of spiritual, commercial, residential, socially promiscuous, intensely contested landscapes of inhabitation. In other words, it was more important to produce settings where daily life required a continuous "updating" of individual sensibility and performance, of what one could be as a shifting composite with others, rather than securing a sense of an integral self.

In the mangle that is Penjaringan and Maystripalya, it is difficult to criticize the ways in which the political organizations of the poor are not exercising their strengthened capacities for climate mitigation objectives. They have increasingly entered into a larger political fray, as candidates for various offices come to seek their endorsement and offer promises of a better future. They position the poor as habitants tending to the environmental surrounds, keeping riverbeds clean, spearheading a local economy that recycles waste, powering their existence from microgrids, and bringing to public attention not so much their rights as the impoverished but their capacities to live sustainably on precarious terrain.

As such, they have managed to secure increased volumes of financial resources, even as these tend to particularize certain areas and actors as residents with enhanced privileges. Although important, there is nothing new about this politics, and it is unclear how it impacts or folds in the long-honed capacities of different domains of Penjaringan and Maystripalya to silently and implicitly extend themselves to each other. This dilemma certainly does not obviate the importance of patient and incremental interventions that attempt to provide more efficient and sustainable urban services to poor residents. It does not obviate the ways in which increasing the territories of operation for the poor—enhanced articulations among their economic activities and practices of caretaking and networking—might enable them to define the dispositions of spatial development away from the avarice of big developers.

Yet many community-level interventions undertaken for climate mitigation tend to close down spaces of interchange, as have been operative in many Global South communities via an imposition of fixity, measurability, and transparency, involving a variety of institutional actors, from consulting firms, NGOs, university-based institutes, think tanks, and even activists. The near obsession with mapping and surveying feeds data-management services that reinforce rationalities based on "evidence-based policy." Coupled with GIS and GPS cartographic systems, land functions are more precisely categorized, recalling colonial practices, and complex social claims to land and resources are formatted into singular property forms—all of which aid and abet the prospects for the future resettlement of the poor regardless of the political settlements now operative.

Such orientations point to a shift away from negotiability to calculation as the essential method for determining the value of life practices. Establishing multiple parameters of efficiency and viability subject to continuous monitoring and assessment replaces processes of adaptability to how specific contexts experiment with their own identifies and functions. Ever more encompassing systems of interoperability that interrelate various data streams, profiles, and histories to determine predilections, probable outcomes, and eligibilities based on an individual's relative standing with an enlarged pool of others determine the viable and worthiness of particular courses of action. Not only is the human as a locus of self-reflection and autonomy being increasingly dispensed with in this process, but it is increasingly converted into a particular status of privilege and eligibility that must be defended not through demonstration but all the more extensively through techno-scientific rationales.

AbdouMaliq Simone is senior professorial fellow at the Urban Institute, University of Sheffield, and visiting professor of urban studies at the African Centre for Cities, University of Cape Town. Recent publications include *Improvised Lives: Rhythms of Endurance for an Urban South* (2018) and *The Surrounds: Urban Life within and beyond Capture* (forthcoming).

Solomon Benjamin is faculty at the Humanities and Social Science Department of the Indian Institute of Technology Madras, India. A member of the Frozen Fish Collective, his interests include practices of land occupancy, transnational economies across India and China, and critical urban studies and art practice.

Notes

1. Shatkin, "Futures of Crisis."
2. Tomasa and Setijadi, "New Forms of Political Activism."
3. Dekeyser, "Pessimism, Futiity, and Extinction."
4. Abidin, Henri, and Gumilar, "Land Subsistence of Jakarta"; van Voorst, "Formal and Informal Flood Governance."
5. Octivianti and Charles, "Disaster Capitalism?"
6. Thompson, "A Dutch Garuda"; Salim, Keith, and Fisher, "Maladaptation on the Waterfront"; Wade, "Hyper-Planning Jakarta."
7. Colven, "Understanding the Allure."
8. Putri and Rahmanti, "Jakarta Waterscape"; Goh, "Urban Waterscapes."
9. Lin and Moon, "Negotiating Time."
10. Putri, *Black Water*; Furlong and Kooy, "Worlding Water Supply"; Kooy, Walter, and Prabaharyaka, "Inclusive Development."
11. Putri, "Sanitizing Jakarta."
12. Kusno, *After the New Order*; Leitner and Sheppard, "From Kampungs to Condos?"
13. Harjoko and Adianto, "Topology and the Web"; Tilley, Elias, and Rethel, "Undoing Ruination in Jakarta."
14. Kusno, *After the New Order*; Kusno, "Power and Time Turning."
15. Budiman, "The Middle Class and Morality Politics."
16. Lobina, Weghmann, and Marwa, "Water Justice."
17. Padawangi, "Reform, Resistance, and Empowerment"; Padawangi, "Building Knowledge."
18. Kusno, "Power and Time Turning."
19. Leitner, Colven, and Sheppard, "Ecological Security."
20. Mulyana, *Decent Work in Jakarta*; Padawangi, "Building Knowledge."
21. Kusno, "The Post-Colonial Unconscious."
22. Tadiar, *Remaindered Life.*
23. Rukmana, "Indonesian Spatial Planning"; Savirani and Aspinall, "Adversarial Linkages."
24. Aljunied, "Sufi Cosmopolitanism"; Benjamin, "The Multilayered Urbanization."
25. Baindur, "Bangalore Lake Story."
26. D'Souza and Nagendra, "Changes in Public Commons"; Patil et al., "The Story of Bengaluru's Peripheries."
27. Goldman, "Speculative Urbanism."

References

Abidin, Hassanudin, Andreas Henri, and Irwin Gumilar. "Land Subsidence of Jakarta (Indonesia) and Its Relation with Urban Development." *Natural Hazards* 59, no. 3 (2011): 1753–71.

Aljunied, Muhd Khairudin. "Sufi Cosmopolitanism and the Subversion of Colonial and Postcolonial Enclosures: The Life and Afterlife of Habib Noh Syed." Paper presented at the conference "Wild Spaces and Islamic Cosmopolitanism in Asia," Asian Research Institute, National University of Singapore, January 14–15, 2015.

Baindur, Meera. "Bangalore Lake Story: Reflections on the Spirit of the Place." *Journal of Cultural Geography* 31, no. 1 (2014): 32–56.

Benjamin, Solomon. "The Multilayered Urbanization of South Canara in Second Part—Land, Society, Belonging." In *Subaltern Urbanisation in India: The Dynamics of Ordinary Towns*, edited by Eric Denis and M. H. Zérah, 199–233. New Delhi: Springer, 2016.

Budiman, Maneke. "The Middle Class and Morality Politics in the Envisioning of the Nation in Post-Suharto Indonesia." *Inter-Asian Cultural Studies* 12, no. 4 (2011): 482–99.

Colven, Emma. "Understanding the Allure of Big Infrastructure: Jakarta's Great Garuda Sea Wall Project." *Water Alternatives* 10, no. 2 (2017): 250–64.

D'Souza, Rohan, and Harini Nagendra. "Changes in Public Commons as a Consequence of Urbanization: The Agara Lake in Bangalore, India." *Environmental Management* 47 (2011): 840–50.

Dekeyser, Thomas. "Pessimism, Futility, and Extinction: An Interview with Eugene Thacker." *Theory, Culture, and Society* 37, nos. 7–8 (2020): 367–81.

Furlong, Kathryn, and Michelle Kooy. "Worlding Water Supply: Thinking beyond the Network in Jakarta. *International Journal of Urban and Regional Research* 41, no. 6: (2017): 888–903.

Goh, Kian. "Urban Waterscapes: The Hydro-Politics of Flooding in a Sinking City." *International Journal of Urban and Regional Research* 43, no. 2 (2019): 250–72.

Goldman Michael. "Speculative Urbanism and the Making of the Next World City." *International Journal of Urban and Regional Research* 35, no. 3 (2011): 555–81.

Harjoko, Triatno Yudo, and Joko Adianto. "Topology and the Web of Informal Economy: Case Study of Kaki Lima and Its Twisted Networks in the Market of Kebayoran Lama, Jakarta. *Asian Journal of Environment-Behaviour Studies* 3, no. 6 (2011): 179–89.

Kooy, Michelle, Carolin Tina Walter, and Indrawan Prabaharyaka. "Inclusive Development of Urban Water Services in Jakarta: The Role of Groundwater." *Habitat International* 73 (2018): 109–18.

Kusno, Abidin. *After the New Order: Space, Politics, and Jakarta*. Honolulu: University of Hawai'i Press, 2013.

Kusno, Abidin. "The Post-Colonial Unconscious: Observing Mega-Imagistic Urban Projects in Asia." In *The Emerging Asian City: Concomitant Urbanities and Urbanisms*, edited by Vinayak Bhame, 168–78. Abingdon, UK: Routledge, 2013.

Kusno, Abidin. "Power and Time Turning: The Capital, the State, and the Kampung in Jakarta." *International Journal of Urban Sciences* 19, no.1 (2015): 53–63.

Leitner, Helga, Emma Colven, and Eric Sheppard. "Ecological Security for Whom? The Politics of Flood Alleviation and Urban Environmental Justice in Jakarta, Indonesia." In *The Routledge Companion to the Environmental Humanities*, edited by Ursula K. Heise, Jon Christensen, and Michelle Niemann, 194–205. London: Routledge, 2017.

Leitner, Helga, and Eric Sheppard. "From Kampungs to Condos? Contested Accumulations through Displacement in Jakarta." *Environment and Planning A: Economy and Space* 50, no. 2 (2017): 437–56.

Lin, Cindy, and Andy Moon. "Negotiating Time: Design as Historical Practice." *PORTAL Journal of Multidisciplinary International Studies* 13, no. 2 (2016): 1–8.

Lobina, Emanuele, Vina Weghmann, and Marwa Marwa. "Water Justice Will Not Be Televised: Moral Advocacy and the Struggle for Transformative Municipalization in Jakarta." *Water Alternatives* 12, no. 2 (2019): 725–48.

Mulyana, Wahyu. *Decent Work in Jakarta: An Integrated Approach. Jakarta, Indonesia.* https://www.ilo.org/wcmsp5/groups/public/---asia/---ro-bangkok/documents/publication/wcms_174991.pdf (accessed January 4, 2018).

Octavianti, Thanti, and Katrina Charles. "Disaster Capitalism? Examining the Politicization of Land Subsidence Crisis in Pushing Jakarta's Seawall Megaproject." *Water Alternatives* 11, no. 2 (2018): 394–420.

Padawangi, Rita. "Climate Change and the North Coast of Jakarta: Environmental Justice and the Social Construction of Space in Urban Poor Communities." In *Urban Areas and Global Climate Change*, Research in Urban Sociology 12, edited by William G. Holt, 321–39. Bingley, UK: Emerald Group, 2012.

Padawangi, Rita. "Reform, Resistance, and Empowerment: Constructing the Public City from the Grassroots in Jakarta, Indonesia." *International Development Planning Review* 36, no. 1 (2014): 33–50.

Padawangi, Rita. "Building Knowledge, Negotiating Expertise: Participatory Water Supply Advocacy and Service in Globalizing Jakarta." *East Asian Science, Technology and Society* 11, no. 1 (2017): 71–90.

Patil, Sheetal, Raghvendra B. Dhanya, S. Vanjari, and Seema Purushothaman. "The Story of Bengaluru's Peripheries: Urbanisation and New Agroecologies." *Economic and Political Weekly* 53, no. 41 (2018): 71–77.

Putri, Prathiwi. "Black Water—Grey Settlements. Domestic Wastewater Management and the Socio-Ecological Dynamics of Jakarta's Kampungs." PhD diss., KU Leuven. https://limo.libis.be/primo-explore/fulldisplay?docid=LIRIAS1953321&context=L&vid=Lirias&search_scope=Lirias&tab=default_tab&lang=en_US&fromSitemap=1.

Putri, Prathiwi. "Sanitizing Jakarta: Decolonizing Planning and Kampung Imaginary." *Planning Perspectives* 34, no. 5 (2019): 805–25.

Putri, Prathiwi, and Aryani Sari Rahmanti. "Jakarta Waterscape: From Structuring Water to 21st Century Hybrid Nature." *Nakhara: Journal of Environment and Design* 6 (2010): 59–74.

Rukmana, Deden. "The Change and Transformation of Indonesian Spatial Planning after Suharto's New Order Regime: The Case of the Jakarta Metropolitan Area." *International Planning Studies* 20, no. 4 (2015): 350–70.

Salim, Wilmar, Keith Bettinger, and Micah Fisher. "Maladaptation on the Waterfront: Jakarta's Growth Coalition and the Great Garuda." *Environment and Urbanization ASIA* 10, no. 1 (2019): 63–80.

Savirani, Amalinda, and Edward Aspinall. "Adversarial Linkages: The Urban Poor and Electoral Politics in Jakarta." *Journal of Current Southeast Asian Affairs* 36, no. 3 (2017): 3–34.

Shatkin, Gavin. "Futures of Crisis, Futures of Urban Political Theory: Flooding in Asian Coastal Megacities." *International Journal of Urban and Regional Research* 43, no. 2 (2019): 207–26.

Tadiar, Neferti X. M. *Remaindered Life*. Durham, NC: Duke University Press, 2022.

Thompson, Rachel. "A Dutch Garuda to Aave Jakarta? Excavating the NCICD Master Plan's Socio-Environmental Conditions of Possibility." In *Claiming Spaces and Rights to the City*, edited by Jorgen Hellman, Maries Thynell, and Roanne van Voorst, 138–56. London: Routledge, 2018.

Tilley, Lisa, Juanita Elias, and Lena Rethel. "Undoing Ruination in Jakarta: The Gendered Remaking of Life on a Wasted Landscape." *International Feminist Journal of Politics* 19, no. 4 (2017): 522–29.

Tomasa, Dirk, and Charlotte Setijadi. "New Forms of Political Activism in Indonesia." *Asian Survey* 58, no. 3 (2018): 557–81.

van Voorst, Roanne. "Formal and Informal Flood Governance in Jakarta, Indonesia." *Habitat International* 52, no. 3 (2016): 5–10.

Wade, Matt. "Hyper-Planning Jakarta: The 'Great Garuda' and Planning the Global Spectacle." *Singapore Journal of Tropical Geography* 40, no. 1 (2019): 158–72.

Energy States

Unequal Burdens, Climate Transitions, and Postextractive Futures

Ashley Dawson and Macarena Gómez-Barris

Shortly after US-backed far-right forces toppled Bolivia's Movement Toward Socialism (MAS) government in 2019, Evo Morales, the country's ousted first Indigenous president, gave an interview in which he said, "My crime, my sin, is to be an Indian, and to have nationalized our natural resources, removed the transnational corporations from the hydrocarbon sector and mining."[1] During his fifteen years in power, Morales and MAS moved decisively to regain national sovereignty over the nation's natural resources, taking control of strategic utilities and industries such as water and electricity as well as extractive industries such as the natural gas sector and mining. MAS reinvested the proceeds of extraction in social programs targeting the country's poor, thereby cutting poverty in half.[2]

One of the foundations of Morales's plan for further economic growth was a "100 percent national" lithium industry. Lithium, an ultralight metal that allows storage of a large amount of energy in a small space, is used in the batteries of smartphones, laptops, and electric vehicles and is critical to emerging clean energy storage technologies. Morales's overthrow has become known as the world's first "lithium coup," an idea no doubt boosted by the consummately arrogant tweet by Elon Musk, CEO of electric vehicle manufacturer Tesla, who proclaimed, "We will coup whoever we want!" shortly after Morales was ousted.[3] Whether Morales's plans to process lithium domestically rather than simply exporting the raw materials to the Global North were a decisive factor in the coup or not, global competition for the rare earth metals that are key to energy storage technologies is undeniably heating up. The World Bank estimates

Social Text 150 • Vol. 40, No. 1 • March 2022
DOI 10.1215/01642472-9495103

that demand for lithium and other rare earth metals will grow 450 percent by 2050 as the world transitions from fossil fuels to renewable energy.[4] Already interimperial tensions are swirling as corporate behemoths like Tesla and powerful nations such as the United States and China vie to control and extract these resources.[5] The sunset of fossil capitalism seems to portend an intensification of extractive geopolitics.[6]

The notion of a Bolivian lithium coup resonated to the extent that it did because the capitalist world system is already massively dependent on exponentially growing rates of extraction. As the United Nations puts it, "The global material footprint rose from 43 billion metric tons in 1990 to 54 billion in 2000, and 92 billion in 2017—an increase of 70 per cent since 2000, and 113 per cent since 1990. Without concerted political action, it is projected to grow to 190 billion metric tons by 2060. What's more, the global material footprint is increasing at a faster rate than both population and economic output. In other words, at the global level, there has been no decoupling of material footprint growth from either population growth or GDP growth."[7]

It takes massive quantities of energy to push such exponentially increasing volumes of stuff through contemporary infrastructures of extraction, production, consumption, and wastage. Today's world ecological system is built on ceaselessly increasing linear flows of materials using ever-intensifying quantities of energy—all on a finite resource base. As energy analysts such as Vaclav Smil have demonstrated, there is a tight correlation between human energy use and gross world product.[8] If, that is, fossil capitalism generated a new energy system fundamentally distinct from preceding systems dependent on and limited by energy derived from muscles, water, and wind, it also catalyzed an intensification of power—the rate at which work is done, materials are extracted and transformed, cities raised and razed. Indeed, ecologist Howard Odum called fossil fuel energy a "power subsidy" and wrote that "the prosperity of some modern cultures stems from the great flux of oil fuel energies pouring through machinery and not from some necessary and virtuous properties of human dedication and political design."[9] This fundamental fact is all too often forgotten.

Philosophers Antti Salminen and Tere Vadén argue in *Energy and Experience* that this leads to very basic confusions: "The neglect of the existential aspect of oil leads to the illusion that capitalism (or socialism) as such leads to economic growth and prosperity. There are no real-life examples of capitalist industrial production producing wealth for large masses without fossil fuels."[10] For Salminen and Vadén, fossil fuels provide a form of unrecognized work that "con-distances," that binds the close and the familiar to distant forces, from the concentrated remains of millennia-old sunshine to the extractive labor of workers in distant lands.[11]

Blithe pronouncements about green capitalism's dematerialization are a perfect instance of such "con-distancing."[12]

In this essay, we challenge these forms of green capitalist mystification using the notion of the energy state, which refers to the structures of "fossil feeling" generated by access to historically unprecedented quantities of energy derived from fossil fuels. By tapping reserves drawn from two-hundred-million-year-old sunlight, those living in the world's rich, imperial nations have over the last two centuries or so experienced unprecedented forms of temporal and spatial acceleration and expansion. The immense power subsidy provided by fossil fuels permits an increasing rate of work that is experienced on an existential level through temporal speedup, as the scale and pace of change gets faster and faster. The ideological form that this foundational drive takes is a devotion to and reification of notions of growth and progress. Cultural critics such as Fredric Jameson termed the resulting state of temporal and geographical disorientation "postmodernism," although the links to extreme extraction of fossil fuels have tended to elude commentators on the postmodern condition.[13]

But we also use the concept of the energy state to refer to the constitution of the modern apparatus of governance by unprecedented flows of energy. Societies with complex governance systems are based on significant energy surpluses, since it takes huge quantities of energy to support such complexity.[14] Timothy Mitchell refers to this often-elided foundation of state formation as carbon democracy, but we prefer the more capacious term "energy state" since it alludes specifically to energy flows rather than to the rather abstract and ubiquitous element carbon and also because it suggests that what is at issue is a certain form of state that is generated by intense extraction and exploitation of fossil energies.[15] Whether it is nominally liberal democratic, democratic socialist, or communist, the energy state is characterized by a fundamental dedication to intensifying rates of energy exploitation and linear material flows. The energy state can have a democratic veneer but, we argue, tends toward forms of inequality and even absolutism. This is not simply, as Mitchell argues, because the particular material qualities of petroleum tend to facilitate centralized control of resource flows, with the attendant forms of inequality that this perpetuates.[16] It is also because the energy state is built on fundamental forms of alienation, including spatial and temporal sprawl, feckless resource expropriation and depletion, and forms of cost externalization that generate sacrifice zones, slow violence, and "surplus populations" targeted for various kinds of premature death.[17] As imperial nations have mined fossil resources, for example, they have been able to subjugate and plunder much of the rest of the planet with the power they have uncorked. They have also colonized the future through their pollution of the atmosphere and the oceans with carbon emissions.

In this essay we discuss two examples of activist struggles around energy infrastructures that challenge and seek to overturn these dynamics of the energy state. As we will see through our discussion of movements in New York City and Quito, Ecuador, energy democracy might be seen as the antithesis of the "extractivist" agenda and, indeed, of the energy state more broadly inasmuch as it is constituted by the efforts of grassroots groups to wrest control of energy away from corporate and/or state-based oligopolies and place them in the hands of the people. Movements for energy democracy strive to shut down fossil capitalist projects and to build decentralized, democratically managed, equitably shared, non-fossil-based energy resources. We explore the strategies used by energy democracy movements in Ecuador as they resisted the modes of state governance that reproduced the hegemonic formation of an energy state, a state powered by nonrenewable natural resources, maximizing profit for development goals by making profits off Indigenous territories in Ecuador's Yasuní region. We juxtapose this fight against extractivism with the work of an environmental justice organization to establish a community solar power cooperative in a predominantly Latinx neighborhood in New York City. As activists in the city fought for this co-op, they found they had to put pressure on city government in order to win the space and material support for their experiment in energy democracy. Energy democracy thus inevitably involved an effort to shift and even transform the state on different scales. Local success in this struggle is tempered by the domination of the city as a whole by the community-smashing, climate-change-inducing prerogatives of the real estate industry.

Our work in this essay is to think hemispherically, since movements for energy democracy and climate justice in the Americas often find ethical and ontological ground in the concept of *el buen vivir* (the principle of good living). At its most fundamental level, this principle refers to the organization of social and ecological life that is based on Afro-Indigenous principles and the transmission of vernacular practices that maintain a deep and respectful relationship to land, place, and the natural world. The notion of *el buen vivir* decenters the human's importance by focusing on how other-than-human life possesses its own sets of rights, logics, and capacities that cannot be solely apprehended, managed, or narrated through human language or scientific technique. Rather than assume knowledge over, or exert a hierarchical relationship to, nature, *el buen vivir* pursues what Atawallpa Oviedo Freire terms "dynamic equilibrium" and "harmony with reciprocity" as ways of relating to the vast nonhuman life all around us.[18] Yet although this grounded principle of *el buen vivir* propels Indigenous movements, communities, relations, and even energy democracy, it has also been used as a rhetoric of state governance to justify an extractivist agenda.

One of the central myths about the state is that progressive and social democratic governments are antithetical to the teleology of capitalist growth. Yet the experience of what we call the energy state in relation to the Latin American Pink Tide wave that took place during the 1990s and 2000s shows otherwise: even though there were important social and economic gains made under progressive governments, they often pursued the dominant model of intensified extractivism. Struggles over the meaning and shape of the energy state, and the definition of *el buen vivir*, are thus constitutive elements of contemporary politics. Similarly, the United States has witnessed intense debates about the shape and scale of energy transition. The intensity of the climate emergency and the need for a rapid and massive energy transition seems to dictate the kind of large-scale intervention that only state-owned and state-managed programs are capable of achieving. Must energy transition therefore take place on a national scale? Are decentralized forms of power such as wind and solar cooperatives compatible with policies for energy transition articulated through the nation-state?

A comparative, hemispheric lens is invaluable in addressing these most challenging questions of the energy transition. The histories of grounded, revolutionary processes with respect to expanding energy democracy in the Americas provide important lessons for social movements and just transition in the belly of the imperial beast. Conversely, these regional movements have important roles to play in establishing hemispheric solidarity with movements for successful energy transition and, to think beyond oil dependency in Latin American nations, challenging the assaults waged by global capital that almost inevitably follow such economic transitions. Admittedly, this juxtaposition of movements at opposite ends of the imperial world system may seem jarring. It is undoubtedly true that *all* citizens of the United States benefit in diverse if often intangible ways from the nation's enduring imperial hegemony, including from traditions of extractivism and political domination linked to the energy state. Residents of the United States need to find ways to power down the energy state in order to give the peoples of the Global South room to escape energy poverty. Notwithstanding these divergences, hemispheric movements for energy democracy have significant related interests. As the work of environmental justice activists and scholars has shown, working-class people and communities of color in cities such as New York are targeted for forms of cost externalization (i.e., toxic industries, polluting waste facilities, asthma-inducing traffic arteries) structurally linked to the forms of slow and immediate violence visited on communities in the Global South. Additionally, comparison of situation and tactics between urban activists fighting fossil capitalism in different parts of the Americas should be seen as an essential part of fostering solidarity in order to proliferate the urban climate insurgency on a global scale.

Connecting movements across the boundaries of nation-states will be key to overcoming fossil capitalism and the trajectory toward planetary ecocide it has put us on.

Black, Indigenous and communities of color, and other urban and rural movements in the hemisphere offer important comparative lessons on how to proceed, and thus we discuss ongoing efforts to delink the state from its complicit reproduction of corporate fossil fuel accumulation. Our premise is that the dominant energy paradigm continues to be locked within a colonial matrix of power.[19] The process of delinking, as Walter Mignolo puts it, and disidentifying, as José Muñoz discusses, is necessary toward a more "sust'āinable" energy paradigm.[20] Yet delinking from the colonial nation, and disidentifying with energy-dominant paradigms is not enough, since we must rapidly move toward a model of radical interdependency that imagines shared governance with Indigenous peoples and radical justice for Black peoples and immigrant communities of color within a demilitarized, deindustrialized, and green-converted model beyond the energy state. Such a vision is a monumental but not wholly impossible challenge, one that is achievable by working through multicoalitional and transversal approaches to energy democracy. In sum, our sustained scholarly findings on unequal systems of extraction and carbon dependency, and on radical social and economic alternatives, point the way toward a postextractive, critical, and hemispheric approach to energy studies.

Decolonizing Energy: YASunidos

Since the 1960s, when the US corporation Texaco discovered underground petroleum reserves in Ecuador, the oil commodity petroleum boom has fundamentally shaped the nation's path of economic dependence. Rather than break with the dominant petroleum model, Rafael Correa's government (2007–17) only expanded the Ecuadoran energy state's dependence upon extractive industries, with devastating social and environmental consequences for Indigenous communities. Importantly, under Correa, new contracts were granted for nonrenewable resources in biodiverse Indigenous territories, consolidating transnational relationships with China, the United States, and Canada. These processes expanded the energy state even as they created new forms of energy democracy to challenge colonial processes of Indigenous dispossession. In biodiverse territories like Yasuní, the compromises of Correa's energy state are clear.

In 1998, UNESCO declared the Yasuní region, which encompasses 982,000 total hectares, a world heritage site worthy of international patrimony. Long considered the most biodiverse region in the world, the Yasuní National Park harbors millions of insect species, hundreds of bird species, and the greatest variety of tree species anywhere on the planet. Moreover, as recent scientific studies have indicated, the Yasuní is not a pure space of untamed wilderness but has maintained its biodiversity precisely because of the ingenuity of Indigenous seed selection, interplanting, and the meticulous cultivation and maintenance of biodiversity over a thousand years of systematic care; forest dwellers that live in the region, including the Waorani, Kichwa, Shuar, and "no contact" populations, have carefully protected and cultivated plant life in ways that support its extension. Ironically, biodiversity is not all the Yasuní has; this land is cursed with an estimated 920 million barrels of below-surface oil reserves that, just as in the surrounding areas, are mapped into oil blocks as commodities by extractive industries.

The 1996 Yasuní agreement originally proposed that the Ecuadoran state would refuse extractivist bidding within its two-hundred-thousand-hectare territory. Instead, it would conserve the region for the Indigenous populations that lived there, in addition to providing resources toward the conservation of biodiversity for future generations. Early in his presidency, Correa rendered the rich biosphere as needing protection from the threat of extractive development, as a site of Indigenous heritage and patrimony, whose costs for leaving oil in the ground could be offset from a network of world actors interested in conservation. Correa agreed to the Yasuní-ITT (Ishipingo-Tambococha-Tiputini) proposal, commonly known as Yasuní-ITT, as an opportunity from outside of Ecuador to prevent the exploitation of oil territories from within, showcasing how Ecuador ultimately became the epitome of how advanced nations continue to prey on resource-rich regions and must be accountable for their extractive practices within a resource-dependent world. Yet, the distance between the energy state's rhetoric and actual conservation of Indigenous territories and biodiverse regions is startling in this case.

In a press release dated April 1, 2007, the day the ITT agreement was adopted, Correa's Ministry of Energy and Mines made the following statement:

> Se aceptó como primera opción la de dejar el crudo represado en tierra, a fin de no afectar un area de extraordinaria biodiversided y no poner en riesgo la existencia de varios pueblos en aislamiento voluntario o pueblos no contactados. Esta medida será considerada siempre y cuando la comunidad internacional entregue al menos la mitad de los recursos que se generarían si se opta por la explotación del petroleo; recursos que require la economía ecuatoriana para su desarollo.
>
> [The government] has agreed to leave the oil reserves within the earth, a result that does not affect the extraordinary biodiversity and doesn't put at risk the existence of various communities in groups of no contact. This

> policy will be considered as long as the international community gives half of the resources that would have been generated had the option been for petroleum exploitation. These are resources that the Ecuadoran economy requires for its development.[21]

To leave the underground oil reserves alone, the Correa government asked the international community to pay $350 million over ten years to the Ecuadoran state, stressing the importance of protecting Pachamama, or Mother Earth. On the one hand, Correa's model of petitioning the ecological conscience of the world to make an investment in Yasuní offered an important alternative for Global South governments, resolving the burden of absorbing preservation costs by externalizing them to the wealthy nations of the North. The Yasuní-ITT proposal provided a plausible, if complex, alternative to existing conservation models through a global system that assumed patronage of regional resources. On the other hand, the wording of the proposal obfuscates the fact that parts of the eastern Ecuadoran Amazon had long been sold off to more than a dozen petroleum corporations. Maps by local activists showed how the region was carved up into "oil blocks," serving as contradictory evidence that these territories would be legally protected by the state.

In August 2013, then-president Correa announced that the state effort to attract international funding for the Yasuní had come up short, and the treaty agreement would be terminated, recusing the state from any further responsibility as protectorate of the region. Given that the Yasuní represents incalculable genetic and species variation, such news represented the failure of legal and political structures to secure ecological futures for highly biodiverse zones. Further, Indigenous movements in Quito and across Ecuador were outraged that despite the commitment to protect ancestral territories, short-term revenues and profit seemed to be still in the background. Petroleum exploitation continued to be an ongoing option for Correa's government.

President Correa's three-year plan, "El Buen Vivir: Plan Nacional, 2013–2017," shows how the concept of plentiful living was used for capital gains.[22] It also reveals the distance between state rhetoric and the policy implementation of the dynamic and mutual reciprocal definition of *el buen vivir* we outlined in the introduction. In the document, the commitment to pursue *el buen vivir* serves as a measure against "infinite development and expansion," taking economic redistribution into consideration, while also extending third-generation social, economic, cultural, and environmental rights. Even though Objective 11 of the state plan argues that Ecuador should consider converting its economic and industrial sectors to align with *el buen vivir* approaches, subsequent parts of the document argue that the model would be incomplete without natural resources to motor economic development. As the plan states:

> El Ecuador tiene una oportunidad histórica para ejercer soberanamente la gestión económica, industrial y cientifica, de sus sectores estratégicos, Esto permitirá generar riqueza y elevar en forma general el nivel de visa de nuestra población. Para el Gobierno de la Revolución Ciudadana, convertir la gestión de los sectores estratégicos en la punta de lanza de la transformación tecnológica e industrial del país, constituye un elemento central de ruptura con el pasado.
>
> Ecuador has the historic opportunity to exercise sovereignty in the economic, industrial and scientific development of its most strategic sectors. This permits the generation of wealth and an elevation of a general level of wealth in the population. For the government of the Citizen's Revolution converting the management of strategic sectors is the point of technological and industrial transformation of the country. This is central to rupturing with the past.[23]

Though there is a will to conserve biodiversity, the focus in this passage is on the nation-state and its historical difficulty in rupturing with dependent sectors of the economy. After all, the "Citizen's Revolution" refers to a redistributive state that would have increased autonomy to condition its own path to capitalist development within a hitherto global order. At the same time, the real emphasis in these passages is on the "technological and industrial transformation of the country," including the infrastructures for mining, hydroelectricity, and petroleum extraction that are implicated in modernizing the nation. Further, there is little indication regarding how the state would accomplish the gigantic task of decolonization, and still less analysis about what sovereignty for Ecuador actually means in relation to Indigenous rights, territorial agreements, and the important work of returning lands to stewardship by First Nations peoples.

One lesson of the Yasuní-ITT treaty is that the energy state can rhetorically advance principles of decolonization, even while pursuing an extractivist agenda that does not reorganize the fundamental exploitative character of market capitalism. For instance, leaked documents revealed secret negotiations between the Ecuadoran state and Chinese petroleum companies over "Bloque 31," which designated the territory of the Yasuní National Park for oil extraction as early as 2009. While Correa initially supported the agreement publicly and forwarded action toward securing protection of the region, by 2009 this had vastly changed through actions that were designated as illegal in the treaty. In this transformation, *el buen vivir*, rather than a method to reckon with the future of conservation, actually allowed for new extractive corridors to be built in eastern Ecuador.

The Earth Constitution

Despite significant debates around questions of "plurinationalisms," including the depth of mining concessions, Indigenous voices over territories, and the extent of ongoing extractivist projects, the revised constitutions of both Ecuador and Bolivia made an enormous step forward by granting rights to Mother Earth (*Pachamama*) and to future generations. In this sense, the 2008 Ecuadoran constitution seemingly shifted the potentiality of the law to regulate urban extraction of biodiverse regions, opening an apparatus for representing nature. For instance, article 71 or the "Rights of Nature" in chapter 7 of the constitution considers the degree to which Pachamama "reproduces and makes life possible":

> La naturaleza o Pacha Mama, donde se reproduce y realiza la vida, tiene derecho a que se respete integralmente su existencia y el mantenimiento y regeneración de sus ciclos vitales, estructura, funciones y procesos evolutivos. Toda persona, comunidad, pueblo o nacionalidad podrá exigir a la autoridad pública el cumplimiento de los derechos de la naturaleza. Para aplicar e interpretar estos derechos se observaran los principios establecidos en la Constitución, en lo que proceda. El Estado incentivará a las personas naturales y jurídicas, y a los colectivos, para que protejan la naturaleza, y promoverá el respeto a todos los elementos que forman un ecosistema.
>
> Nature or Pachamama, where life is reproduced and made possible, has the right to exist, persist, maintain and regenerate its vital cycles, structure, functions and evolutionary processes. Every person, people, community or nationality, will be able to demand the recognitions of rights for nature before the public authority. The application and interpretation of these rights will follow the related principles that are established in the Constitution. The State will incentivize natural rights and juridical persons as well as collectives to protect nature; it will promote respect towards all the elements that form an ecosystem.[24]

From its first article, the language of the constitution links Indigenous concepts of sovereignty to Earth rights, stating that the nation is a "unitary, intercultural, multinational and secular State" that is governed using a decentralized approach. Moreover, the constitution specifies the modes through which this decentralized form of governance should be exercised:

> Sovereignty lies with the people, whose will is the basis of all authority, and it is exercised through public bodies using direct participatory forms of government as provided for by the Constitution. Nonrenewable natural

> resources of the State's territory belong to its inalienable and absolute assets, which are not subject to a statute of limitations.[25]

The language of sovereignty describes the protections by the law to the natural world, even as it has not prevented extractivist corporations from entering into protected terrains. The only way that these legal protections are guaranteed is through the powerful coalitional forces of Indigenous movements with rural and urban allies that produce a network of social actors to both ensure actualization of Earth rights protections, as well as ongoing forms of participatory democracy that force accountability from the energy state.

YASunidos

The consequences of Correa's government have been to fragment the environmental and Indigenous movements that put him in power in the first place, even as in the current period a range of Indigenous activisms, youth leaders, eco-feminist and socialist urban and rural alliances continue to carry out an antiextractivist vision of the future. Another lesson of the energy state is its capacity to criminalize and denounce those that challenge its market-oriented and petroleum-extractive agenda.

When Correa announced in 2013 that the Yasuní-ITT treaty would not be upheld, activists protested how the region had seemingly moved so rapidly from its protected status as a site of conservation to an oil block tracked for petroleum extraction. In the days that followed, hundreds of people across the nation gathered in vigils, shattered by the announcement of this repeal, and by Rafael Correa's apparent turning away from the postdevelopment and *el buen vivir* paradigm. As Diana Coryat describes, such gatherings spawned the YASunidos movement, led by urban youth who had come out of genealogies of ecological activism, media collectives, and horizontal organizations rejecting the extractivist model.[26]

For ecologically minded youth, the Yasuní represents a mental and physical sanctuary, a psychological space of respite from a climate change world, a place where biodiversity could thrive amidst an increasingly dystopic national context of mega-development projects. Thus, the subsequent shock of finding out that Correa ruptured the agreement, when a road was constructed as an extractive corridor through the Yasuní-ITT zone, produced an outraged response by the YASunidos coalition with effects that continue to reverberate. As a coalitional movement comprised of artists, Indigenous activists, journalists, middle-class ecologists, and land defenders, YASunidos social mobilizations create links between rural, urban, and Indigenous movements to produce new transversal and coalitional forms of antiextractive organizing. Yasuní functions as a powerful collective symbol of the global devaluing of Indigenous territories and biodiverse ecologies, helping to grow the coalition in powerful ways.

In the following partial statement by the YASunidos, one can access the extent to which a different consciousness moves the coalition, through a call that also specifies central principles of *el buen vivir*. In 2014, YASunidos powerfully addressed the importance of leaving oil in the ground "indefinitely" to support the sovereignty of Indigenous peoples as guaranteed by article 57 of the constitution. As the statement says:

> We demand that our Ancestral, Natural Heritage is not sacrificed and opt for post-oil alternatives. We support a truthful and transparent debate about our economic model and our energy base. Also, we demand that the government let us show our disagreement through the legitimate exercise of protest without repression and criminalization. . . .
>
> Just like the Yasuni-ITT Initiative, we propose a search for alternatives, we propose breaking away from the schemes with courage, in short, we propose a social revolution that challenges the values of energy consumption and that prioritizes the common good, defending the idea of the "Good Living." . . .
>
> We are aware that more than one person has attempted to use our platform to their own advantage, which is why it is necessary to clarify that we fight for life and an alternative to the extractivist model. We are citizens, not only urban citizens, and we are aware of the disasters that oil extraction generates for nature, humanity and the economy. With a strong belief that this is the moment to take the debate to the streets with the participation of everyone, we hope to overcome the oil dependency imposed on us, that moreover further aggravates global warming, environmental destruction, puts the lives of peoples in voluntary isolation at risk and threatens not only the future of Ecuadorians but also that of humanity.[27]

YASunidos' search for alternatives that break "away from the schemes with courage" and propose "a social revolution that challenges the values of energy consumption, and that prioritizes the common good, defending the idea of 'Good Living'" is at the core of the YASunidos movement and its search for ecological justice and futures. In the statement, redistribution rather than expanded drilling and extraction restages the core of a postdevelopment agenda, connecting the threat of extractivism to the negative balance of global warming.

At the end of the statement, YASunidos makes an invitation to all those "who love the country and who want to collaborate and contribute, to join the movement and walk with us in order to build a better future, as Ecuadorans we say, 'We can!'" While the appeal to nationalism is strong in this closing statement, the fusing of ecological living, postdevelopment, and *el buen vivir* become the central tenets of a radical challenge to the extractivist model and the normativity of global capitalism. Ecological futures, in the case of the YASunidos, are the other side of the Ecuadoran state's extractivist development agenda, where the natural world has not been completely mapped by extractive capitalism, yet retains value precisely because of the diversity of life forms that exist within the forest ecologies. Indeed, one question that biologist and antiextractivist activist William Sacher pursued in my interview with him is "what life forms will continue to exist in the future if we continue to devalue and eliminate biodiversity?"[28]

After the rise of YASunidos in 2013, the state initiated a strong-arm practice of criminalizing ecological activisms, Indigenous territorial defenders, and anyone it deemed as challenging the breaking of the Yasuní-ITT agreement.[29] With the rhetoric of democracy and the importance of the constituent assemblies as backdrop, the rupture of the Yasuní agreement marked a turn toward increased authoritarianism, as the state began using surveillance practices to constrain the activities of YASunidos. Diane Coryat importantly documents how much of this criminalization process was initiated through state-controlled media formats. As she says, "As for members of *Yasunidos*, Correa characterized them in different ways. They were either manipulated by politicians, 'the same stone-throwers as always,' (associating them with a radical left political party that was frequently a subject of scorn on the *Enlaces*), or middle-class urbanites with full bellies that had never been to the Amazon, nor knew what it meant to live without basic services."[30] Correa trivialized ecological activisms, criminalizing land and water defenders, precisely during the period when mega-project contracts expanded their purview toward a new hegemony of extractive contracts and designs.[31]

The important rural and urban linkages that facilitated the YASunidos movements allowed for a way of disrupting state and corporate extractive agendas that even filtered into the ITT proposal itself. According to Acción Ecológica (Ecological Action), a radical independent environmental organization in Ecuador, the marking out of the ITT was organized in complicity with the big transnationals that have oil concessions in the Yasuní I Biosphere Reserve. These companies have provided maps, information, and infrastructure, as well as exerted considerable pressure.[32] In such highly capitalized conditions, where the political stakes of conservation in rural areas sometimes feed the ability to generate new extractive corridors, urban-based coalitions are important to holding energy states accountable. They provide forms of energy democracy that watch over the cozy alliances between corporate-state and transnational actors intent upon dispossession as a means to expand new markets for extractive capitalism.

Transversal land and water defense movements across rural and urban geographies, then, become an important response to these highly capitalized efforts. Coalitional efforts including direct action, grassroots media, and forms of autonomous participation of the YASunidos movement sustain biodiversity through acts of radical land defense. But these efforts also include legal action. For instance, by petitioning the case to the Inter-American Court, YASunidos's work did not constitute a failed effort, as Correa's government seemed to suggest at the time, but rather sparked ongoing strategies and approaches that challenge the paradigm of the energy state.

YASunidos represents an enlivened model of *el buen vivir* offering new opportunities for social ecologies that build fresh forms of democracy beyond the logics and policies of the dispossessing energy state. These lifeways and sources of ecological living do not only challenge the vertical and corporate model of petro-extraction; they are its antidote. We turn now to another example of popular contestation of the energy state, this time from within the largest city in the imperial core.

Power to the People

In fall 2019, the first panels of Sunset Park Solar were installed on the roof of the Brooklyn Army Terminal, and the site became New York City's first cooperatively owned solar garden. Once set up, Sunset Park Solar's nearly two-acre arrays began feeding power back into New York City's energy grid. The energy that Sunset Park Solar pumps into the grid is converted into credits that reduce energy bills for local individuals and businesses who join the co-op. Sunset Park Solar is a project of UPROSE, Brooklyn's oldest Latinx community-based organization and a leading member of New York City's Environmental Justice Alliance. For UPROSE Executive Director Elizabeth Yeampierre, Sunset Park Solar is linked to broader efforts on the part of the climate justice movement to "operationalize just transitions."[33] Solar cooperatives established by environmental justice organizations such as UPROSE thus aim to move predominantly working class, people-of-color neighborhoods like Sunset Park "away from fossil fuel extraction to regenerative energy . . . to make it possible for our communities to start moving off the grid and to start creating mechanisms that help them thrive in the face of climate change."[34]

Making this shift requires tackling what Patricia Yaeger called the energy unconscious, the often intangible but nevertheless powerfully affectual forms of energy and its infrastructures.[35] If the Afro-diasporic somatic power harnessed through slavery fueled the global economic circuits that helped produce the modern energy unconscious, Black and Brown bodies and lives have been—and continue to be—systematically exploited, denigrated, and targeted for destruction by modern energy regimes. As Myles Lennon puts it, "The industrialized transformation of matter was predicated on making black lives not matter."[36] UPROSE intends to reverse this "de-mattering" by helping communities that have been historically exploited by and subordinated to diverse energy states to take power back into their own hands.

This transformation of energy systems is as much ideological as it is material. According to Lourdes Pérez-Medina, UPROSE's climate justice policy and programs coordinator, the Sunset Park Solar co-op helps the community visualize and take ownership of urban energy infrastructures that are normally invisible. Being able to see power in this way is especially salient to communities of color, since they are disproportionately located near and negatively affected by the generation facilities that provide electricity in cities such as New York. While many consumers take the electricity produced by investor-owned utilities such as New York's Con Edison for granted, residents of the city's predominantly Black and Brown neighborhoods suffer disproportionately high rates of asthma, cancer, and heart disease as a result of their proximity to power plants and other toxic facilities such as waste transfer stations, bus terminals, and noxious industrial sites.[37] To make matters worse, low-income communities and communities of color spend a large portion of their income on New York's costly utility charges. According to a Move On survey, New York is the fifth most expensive state for utility rates. These high charges, plus the exorbitant cost of housing, ensure that many families in New York struggle to pay their electric bills each month.[38] It is consequently particularly important for such communities to be able to see—and to look beyond—existing industrial energy infrastructure in the city. As Pérez-Medina puts it, "When you have the experience of going through a community process, seeing years of work go into building something that lives in your community and that you get to benefit from, that's completely different from just turning on the light. You get to own your community amenities and at the same time understand how infrastructure decisions impact our environment. It's very important to make the just transition part of our daily experience and a collective exercise."[39]

This is the meaning of energy democracy: rather than simply stressing decarbonization of the grid, the movement for energy democracy aims to challenge who controls the energy system and who benefits from that system. This is why the struggle for energy democracy isn't simply about requiring a certain percentage of renewable energy in the grid. Although so-called renewable portfolio standards that require energy utilities to produce a specified amount of the electricity they furnish to the public from renewable sources are certainly important and to be supported, simply getting more solar and wind energy into the grid will not be enough to rectify the many forms of exploitation and marginalization that working-class communities of color are subjected to by current energy systems. For example, at present so-called public utilities ensure that wealthy investors benefit from high utility rates, but energy democracy activists want to change this. As the activists Denise Fairchild and Al Weinrub argue, "Energy democracy is a way to frame the international struggle of working people, low-income communities, and communities of color to take control of energy resources from the energy establishment and use those resources to empower their communities—literally (providing energy), economically, and politically."[40]

UPROSE and similar organizations around New York City are constructing what anthropologist Dominic Boyer has called "revolutionary infrastructure."[41] Insofar as infrastructures such as the pipeline and the electric grid store and distribute energy, Boyer argues, they are repositories of potential energy. Drawing on Karl Marx's *Grundrisse*, Boyer suggests that this potential power is not an autonomous force; its productivity "depends wholly on the gelatinized combination of expertise, activity, materials, and forces that have filled the mold of its infrastructural design."[42] Infrastructure thus coagulates the productive powers of labor and its transformation of the natural world in a way that allows this power to be released subsequently as if it were autonomous of labor power—indeed, as commons theorist George Caffentzis argues, in a way that allows capital to chart technological paths to repression.[43]

But the potential energy embedded in infrastructure can also be appropriated to revolutionary ends. It can become what the theorists of autonomy call "constituent power," or the power of the people, although in this case *power* has the double meaning of political and energetic power.[44] The power of the people thus aims at the recapture and realization of the transformative political potential physically embedded in the infrastructures of industrial energy. The solar co-op makes such revolutionary power particularly palpable. While few members of the general public have much grasp of the details of their electricity bills such as the kilowatt hour, simply standing outside in the sunshine gives one an immediate sense of the abundant energy that the sun brings to the Earth every second of the day. For Hermann Scheer, one of the principal architects of Germany's energy transition, solar energy—whether in its direct form or in indirect forms such as wind and biomass—has the physical advantage of ubiquity and superabundance, qualities that permit more efficient and decentralized supply chains that are also conducive to more democratic forms of control.[45] UPROSE's solar co-op aims to harness this democratizing potential of solar radiation for the benefit of the community in Sunset Park.

The idea of revolutionary energy infrastructure may seem intangible, particularly in a US context where people have been habituated to the privatization of energy (and so much else) for over a century.

Nonetheless, for NAACP Environmental and Climate Justice Program Director Jacqui Patterson, renewable energy infrastructure can empower communities that have been historically exploited and marginalized by fossil capitalism, allowing them to generate tangible community wealth.[46] Indeed, UPROSE's Elizabeth Yeampierre stresses that Sunset Park Solar is intended to function as an economic driver for her community. Net metering laws allow solar co-ops to feed power back into the grid, running utility electric meters backward and generating income for the community.[47] In addition, the solar co-op will be linked to an UPROSE-initiated jobs program that will train half of the people installing the solar array. But Yeampierre's vision for renewable power in Sunset Park goes beyond these immediate benefits of the solar co-op. As she explains, "A lot of our environmental victories have been used by developers to promote displacement."[48] The solution to this green gentrification for Yeampierre is a broader mobilization of the community around just energy transition.[49] This means that instead of seeing local blue-collar businesses as an eyesore to be shut down—as well-heeled gentrifiers often do—UPROSE activists like Yeampierre argue that "they can be retrofitted, repowered, made adaptable so that the workers are safer and healthier. We see them as part of the solution."[50] New York State, for example, has commissioned the country's largest offshore wind farm off the coast of Long Island. Instead of assembling the wind turbines in Denmark and shipping them across the Atlantic, Yeampierre argues that the state should use existing industrial waterfront facilities in places like Sunset Park to build the turbines, benefitting the entire community while also diminishing carbon emissions.[51] UPROSE's struggle to turn New York's industrial waterfront into a hub for the manufacturing of renewable energy infrastructure suggests that the just transition component of Alexandria Ocasio-Cortez's Green New Deal is not particularly novel, but rather takes to a national level the long-standing fights of low-income communities and communities of color against both environmental injustice and displacement in today's extreme cities.[52]

These ambitions for climate justice in Sunset Park find their fullest expression in the redevelopment plan for a Green Resilient Industrial District (GRID), created by the Collective for Community, Culture, and the Environment at the behest of UPROSE and the Protect Our Waterfront Alliance (POWWA). The GRID proposal is an effort to preempt rezoning of the waterfront by the city and a group of prominent NYC real estate developers. With surrounding neighborhoods already significantly gentrified, Sunset Park's industrial waterfront—the largest of the city's Significant Maritime and Industrial Areas (SMIA)—is seen as a prime site for commercial transformation. The industrial areas in Sunset Park currently provide twelve thousand jobs, but this employment is in construction, food, and green industrial businesses such as Sims Municipal Recycling Center, businesses that provide jobs for a local workforce likely to be displaced by a greater influx of high-tech, design, entertainment, and big-box retail, as well as the market pressures associated with gentrification. Legitimated using the jargon of "innovation," the city's planned rezoning would in fact support the growth of the kinds of chain stores that are already widespread in Brooklyn and throughout the rest of New York City. The city rezoning plans do nothing tangible to help support a just green transition.

The ugly truth is that New York City has been expanding into flood zones for decades in a manner that is totally irrational from an environmental perspective. The city has no authoritative planning process but instead transforms itself through the kind of piecemeal rezoning that the plans for Sunset Park so well encapsulate. Although rezoning takes place on a local level, within particular neighborhoods in specific boroughs, it nevertheless has a systemic, citywide impact. Under Mayor Michael Bloomberg, for example, fully a third of the city was rezoned—and redeveloped by corporate real estate developers. Much of this redevelopment was in waterfront areas such as the city's other Significant Maritime and Industrial Areas, on land long occupied predominantly by working-class communities of color. The "luxury condos" and high-end retail centers constructed in such areas displaced masses of people in a form of slow and relatively invisible violence that nevertheless must be seen as a complementary assault to the brutal policies of "stop and frisk" that placed communities of color under siege during the Bloomberg era. It can give members of these displaced communities little solace that the sea is coming, sooner or later, for these luxury condos and their well-heeled residents.

In contrast to such unsustainable and elitist policies of rezoning, UPROSE and its partners propose to leverage the existing green and industrial resources and public investments in Sunset Park's waterfront zone. The Green Resilient Industrial District plan integrates climate adaptation and resiliency measures while also providing good local jobs and workforce training for community members. Examples of such green jobs include training in green building trades, the construction of turbines for offshore wind farms, and vehicle electrification. If the city agrees to adopt GRID, it will become a model example of local grassroots-driven planning and implementation of a just transition economy. GRID would provide an incredibly resonant local example of the kind of Green New Deal for which advocates have been unsuccessfully fighting on a national scale. In addition, successful adoption of GRID would signal that the city has decided to side—at least for once—with popular planning rather than with the real estate state.[53]

Here, however, we must return to the question of the energy state, for if UPROSE seeks to install decentralized forms of solar energy infrastructure and thereby to win community empowerment, it is not animated by a libertarian imaginary of delinking entirely from the grid. Organizations such as UPROSE need to continue to engage the state, not simply to win the kind of ambitious just transition programs Yeampierre envisages, but for very basic reasons related to existing structures of social inequality on urban terrain. As her colleague Lourdes Pérez-Medina explains, since many working-class people in Sunset Park rent rather than own their property, fundamental questions about where the co-op's solar panels will actually be installed inevitably arise. "For community-based organizations to be able to tackle this kind of project," Pérez-Medina argues, "they need site control. In a city like New York, which is very real estate heavy, that is amazingly difficult. So, city-owned structures and city properties become a huge asset in this type of development."[54] That is, given the extreme inequalities of urban terrain organized around financialized accumulation, activist organizations often must turn to relatively accessible municipal facilities when seeking space for the arrays that power solar co-ops.

While fully aware of the ways that state power continues to be used against their communities, not just by the NYPD but also by apparently more benign entities such as the New York City Economic Development Corporation, activists from organizations such as UPROSE do not see the state as a monolithic entity. Instead, they seek to intervene in and modify what radical theorist Nicos Poulantzas called the "relation of forces within the state."[55] For Poulantzas, the struggle for socialism consisted in "the spreading, development, reinforcement, coordination, and direction of those diffuse centers of resistance which the masses always possess within the state networks."[56] As Gianpaolo Baocchi has demonstrated, such strategies of mobilizing outside, alongside, and inside the state have characterized movements in the Pink Tide countries of Latin America over the last several decades.[57] Another way of thinking of this may be to employ Pierre Bourdieu's analogy in *Counterfire:* the state has two arms, one punitive, violent, and responsive to the most reactionary segments of capital and the petty bourgeoisie, and the other redistributive, disciplinary, and subject to circumscribed but undeniable inroads by long histories of struggle by working-class people and people of color.[58] While both arms of the state ultimately serve the interest of capital, the point made by theorists such as Baocchi, Bourdieu, and Poulantzas is that the autonomous power of the people must not simply put pressure on the state but must also seek to leverage those "centers of resistance" that exist within the state. Poulantzas's arguments about the variegated character of the state and the importance of mobilizing around sectors and scales where community power has gained some purchase are particularly relevant in relation to the contemporary energy state.

As an illustration of the importance of these theorizations of state power in general on the specific terrain of the energy state, consider the spread of community solar in the United States today. Less than half of US community solar projects have any participation from low-income households; of these, only about 5 percent include a sizable share, or more than 10 percent.[59] Under pressure from the climate justice movement, however, twelve states and the District of Columbia have developed a series of mandates, financial incentives, and pilot programs to help low-income communities access shared solar. These initiatives are transformative: when fully rolled out, they will impact about fifty million households, or 44 percent of the population in the United States.[60] They will bring all of the benefits described by the UPROSE activists, including collective access, economic empowerment, and community control. Gaining access to these empowering benefits of community solar requires engagement with and pressure on the state, whether on an urban, state, or federal scale. But it also requires figuring out how to construct energy infrastructures that can exist autonomously from the grid, while also at times pumping power back into that grid in order to access the resulting credits. In sum, community solar power must deploy a politics that exists "in-against-and-beyond the state," in the words of energy democracy activist and scholar James Angel.[61] Rather than cultivating imaginaries of complete energy autonomy, we might instead see UPROSE generating what we would call "energetic disidentifications."[62] Drawing on the work of José Esteban Muñoz, we suggest that energetic disidentifications entail the construction of energy infrastructures and communities that "neither opt to assimilate within a structure nor strictly oppose it; rather, disidentification is a strategy that works on and against dominant ideology"[63]—and, we would add, also on and against existing energy infrastructures, helping shift the relation of forces within the energy state.

While it is important to celebrate the vision and successes of movements for energy democracy in the United States, it is important to remain aware of the significant obstacles they confront. UPROSE, Soulardarity in Detroit,[64] Lakota Solar Enterprises on the Pine Ridge Reservation,[65] and allied organizations for energy democracy all struggle to establish community power within the world's foremost imperial energy state. Under the Barack Obama and Donald Trump regimes, the fracking revolution turned the United States into a fossil capitalist juggernaut, one that sought to export gas and oil while maintaining existing fossil production through massive state subsidies and a globe-girdling military-industrial-petroleum complex.[66] The Trump administration joined red states in seeking to criminalize antipipeline demonstrations such as the protest at Standing Rock.[67] Even in nominally progressive states like New York, whose governor signed off on a plan to transition the state to carbon-free electricity by 2040, modern renewables such as solar and wind power still generate a miniscule proportion of electricity.[68] In New York City, for example, only 1 percent of households are powered by solar.[69] All of this underlines the massive inertia and odds represented by the US racial energy state. But it also suggests the outsize importance of the struggle of an organization such as UPROSE for energetic disidentification.

Furthermore, it is important not to fetishize the organizational form of the co-op. This is a particular danger given the popularity of "horizontalism" and commons-based organizational initiatives on the left today.[70] Electric cooperatives admittedly have a distinguished history in the United States but are far more extensive in rural areas than in cities. This is a product of the history of electrification in the early twentieth century, when for-profit companies grew quickly in the nation's cities, providing power for hundreds of thousands of private and commercial customers from centralized plants. Money was to be made for such so-called utilities through economies of scale: the more customers they could get, and the more electricity those customers consumed, the more money they stood to make. But since it was expensive to set up transmission infrastructure across the country's vast rural expanses, nine out of ten of the nation's rural homes remained without electricity as late as the mid-1930s. To rectify this omission of the nation's farmers from what was seen as key elements of modernity, President Franklin Roosevelt established the Rural Electrification Administration in 1935. Although the investor-owned utilities were almost totally uninterested in government subsidies for rural electrification, applications for government loans poured in from farmer-based cooperatives. These co-ops made rural electrification a reality. Today, the nation's nine hundred or so rural electric co-ops serve 13 percent of the population and span three quarters of its land.

All too often the potential of direct democratic control originally embodied in these rural co-ops has been blunted or even totally hijacked. According to research conducted by the Institute for Local Self-Reliance, more than 70 percent of co-ops have voter turnouts of less than 10 percent.[71] But these low turnouts are not—as one might initially guess—a result of voter apathy. Instead, as Benita Wells of the grassroots organization One Voice, based in Jackson, Missouri, told me, cooperative boards populated and controlled by members of the white good-old-boy Southern hierarchy go to great lengths to disenfranchise the predominantly poor and Black members of the co-op.[72] Although more than 35 percent of the people in Mississippi are African American, fewer than 10 percent of the governing board members of electric cooperatives in the state are Black.[73] Similarly, women make up half the population yet only hold 4 percent of the board seats. Awareness of patterns of suppression of voter turnout, abuse of power, and gouging of poor members with outrageously high rates, the Co-op Democracy Project organized members to challenge the entrenched power of the white hierarchy on co-op boards. Although they had some success, today the racist composition of co-op boards and the political process that eviscerates the democratic potential of the co-ops remains largely unchanged.[74] Not only does this lead to rampant economic exploitation, but in today's context of a coronavirus-fueled economic crash, it can also lead to power cutoffs for some of society's most vulnerable.[75] In addition, these corrupt co-ops show no sign of shifting away from their highly polluting fossil-fuel-based power plants, a situation that shows utter fecklessness in the face of the increasing climate chaos to which their members are highly vulnerable.

Our discussion of UPROSE in New York City has shown the degree to which energy democracy movements must struggle in, with, and beyond the state to win a decarbonized future. And the energy state also must account for its complicity in the carbon web, that is, its responsibility for incalculable loss in relation to biodiversity and disproportionate racialized suffering. Activists currently fighting for a Green New Deal in the United States also need to remember the broader hemispheric interconnections of waste and loss in relation to petro-extraction. New energy paradigms provide alternatives that are not only imaginary but also materialize a series of demands about the future through urgent campaigns for energy democracy.

Conclusion

We have underscored the importance of the complex dynamic between the energy state and popular social movements, focusing on the experiences of the polluting and environmental effects of those at both ends of the production and consumption cycle. One conclusion to draw is the importance of applying the principles of *buen vivir* and mutual exchange by focusing on local energy production or forms of alternative energy that do not externalize costs to Indigenous, Black, and immigrant communities of color. Any Green New Deal or other alternatives to the energy state launched within the United States must include strenuous forms of reparations to make whole nations and peoples who have been subjected to petro-empire across the Americas.[76] And, the Green New Deal must ultimately be global, ensuring that countries not only have adequate economic resources for adaptation to damages caused by the climate crisis but also adopt renewable energy rather than fossil-fueled pathways to address long-standing poverty and inequality.

It has become commonplace to describe the "resource burden" of particularly "well-endowed" geographies of the Global South, a racialized and gendered terminology that presumes that biodiverse and petrol rich territories are there for the normative taking by the wealthiest sectors of the Global North. While we refuse such terminology, we also note the overlapping maps of rising rates of ecocide, genocide, and femicide that remind us that even in the very recent wave of territorial accumulation, the use of state violence by militarized and police apparatuses intensifies, specifically targeting racialized and gendered bodies in areas of high biodiversity and commodity conversion. In terms of historic energy trends in oil dependency, sites of current resource peril and destruction are those that have been also been continually plundered and deforested by systems of franchise, settler, and corporate colonialism. Global capitalist forces and their extreme carbon emission output have dramatically reduced biodiversity, and the energy state has often accelerated, rather than mitigated, these outcomes. By exposing the weaknesses of petro-states, and focusing on new alternatives, we intend to strike back against the deep-rooted forms of melancholia and even despair that petroculture tends to induce.[77]

The current uneven and expanding planetary condition of environmental degradation and the acceleration of climate change suggests that suffering will continue to be acute and disproportional. Under these changing conditions, what is the nature and role of what we have referred to as the "energy state," or the main actor that has perpetuated carbon dependency and the main culprit of climate change? Vast land grabs in the Americas and across the planet have been facilitated by the state, and they have diminished the capacities for human and nonhuman life. After Hurricanes Katrina, Sandy, and María, we witnessed the complicity of the racial and colonial state in its deliberately slow and violent response to racialized suffering. These "weather events" point to a proliferating structure of planetary dystopia demarcated by specific time-space coordinates, or what Kathryn Yusoff has recently termed a billion Black Anthropocenes.[78] In the morass of accumulating narratives about how the apocalypse closes in on us, some have the luxury of assuming the future is impossible. From the position of Latin America, however, and from communities of color within the United States, it has become clearer that racial and extractive capitalism depends upon this collective eco-depression and physical, mental, and spiritual exhaustion in order to perpetuate multispecies extinction. What do alternative energy futures look like that do not reproduce the white settler, heteropatriarchal, and nativist xenophobic tropes of "clean" and "pure" societies? How can we imagine a transition to postextractive and alternative energy economies that account for the weight of colonial and imperial legacies? What kind of energy networks are possible that actually point us toward decolonial futures rather than reinscribe us into the logics of the racial, settler, extractive, and real estate state? In the wasteland of a billion Anthropocenes, one thing is clear: if there is to be a "human" future, it will not be centered on the energy state.

Ashley Dawson is professor of English at the Graduate Center and at the College of Staten Island, City University of New York. He currently works in the fields of environmental humanities and postcolonial ecocriticism. He is the author of three recent books relating to these fields: *People's Power* (2020), *Extreme Cities* (2017), and *Extinction* (2016).

Macarena Gómez-Barris is professor of social science and cultural studies at Pratt Institute, Brooklyn. She works in the fields of decolonial studies, extractivism, and studies of ocean and land with attention to Indigenous issues. She is author of *The Extractive Zone* (2017), *Beyond the Pink Tide* (2018), *Where Memory Dwells* (2009), and *At the Sea's Edge* (forthcoming).

Notes

1. Greenwald, "Interview with Bolivia's Evo Morales."
2. Harasim, "Bolivia's Lithium Coup."
3. Telesur, "Elon Musk Confesses"
4. Harasim, "Bolivia's Lithium Coup."
5. Pitron, *The Rare Metals War.*
6. Gómez-Barris, *The Extractive Zone.*
7. United Nations, "SDG Indicators."
8. Smil, *Energy in Nature and Society*, 397.
9. Odum, *Environment, Power, and Society*, 6.
10. Salminen and Vadén, *Energy and Experience*, 28.
11. Salminen and Vadén, *Energy and Experience*, 21.
12. See, for example, McAfee, *More from Less.*
13. For typical examples of critique that ignores the link between globalized neoliberalism and fossil capitalism, see Jameson, *Postmodernism*, and Harvey, *The Condition of Postmodernity.*
14. Tainter, *The Collapse of Complex Societies.*
15. Mitchell, *Carbon Democracy.*
16. Mitchell, *Carbon Democracy*, 38.
17. Gilmore, *Golden Gulag*, 247.
18. Oveido Freire, *Buen vivir.*
19. See Quijano, "Coloniality of Power, Eurocentrism, and Latin America," 533–80.
20. We want to put pressure on the term "sustainability" here and thus call attention to the terminology of "sust'āinable alternatives to heteropatriarchal systems rooted in genocide, dispossession, and extractive capitalism" following Indigenous Kānaka Maoli creative praxis and concept work.
21. Martinez, "Yasuní El tortuoso camino de Kioto a Quito," 15.
22. Secretaría Nacional de Planificación y Desarrollo, *Buen Vivir.*
23. *Buen Vivir*, 313.
24. "Rights of Nature."
25. Translated from Spanish original into English at Political Database of the Americas, "Constitution of the Republic of Ecuador." To view the entire Ecuadorian Constitution, see Asamblea Nacional, "Ecuadorian Constitution."
26. Coryat, "Extractive Politics."
27. YASunidos International, "YASunidos Manifesto."
28. Sacher, interview.

29. Acción Ecologica, in addition to a number of alternative journalistic efforts, has worked to document the criminalization of indigenous peoples and ecological allies. Despite the Ecuadoran constitution's legal protections of land defenders, surveillance has expanded as discussed in the introduction.

30. Coryat, "Extractive Politics," 8.

31. Gómez-Barris, "Review."

32. See Gómez-Barris, *The Extractive Zone.*

33. Quoted in Pérez-Medina and Yeampierre, "The People's Power."

34. Pérez-Medina and Yeampierre, "The People's Power."

35. Yaeger, "Editor's Column," 309.

36. Lennon, "Decolonizing Energy," 24.

37. Bullard, "Environmental Justice," 151–71.

38. Holmes, "The Cost of Utilities."

39. Pérez-Medina and Yeampierre, "The People's Power."

40. Fairchild and Weinrub, "Introduction," 6.

41. Boyer, "Infrastructure," 223–44.

42. Boyer, "Infrastructure."

43. Caffentzis, "A Discourse on Prophetic Method," 63.

44. Ruivenkamp and Hilton, "Introduction," 8.

45. Scheer, *The Solar Economy*, 89.

46. Cited in Lennon, "Decolonizing Energy," 21.

47. It is worth noting, however, that net metering laws have been replaced by a far less beneficial arrangement called Value of Distributed Energy Resources (VDER) in New York State. On VDER and social justice, see McMullen-Laird, "End of 'Net Metering.'"

48. Pérez-Medina and Yeampierre, "The People's Power."

49. On green gentrification, see Checker, "Wiped Out by the Greenwave," 210–29.

50. Pérez-Medina and Yeampierre, "The People's Power."

51. Similar arguments about resuscitating light manufacturing in the city were made by Mike Wallace based on conversations with activists in the city. See Wallace, *A New Deal for New York.*

52. On the twin economic and environmental perils in contemporary cities, see Dawson, *Extreme Cities.*

53. Stein, *Capital City.*

54. Pérez-Medina and Yeampierre, "The People's Power."

55. Poulantzas, *State, Power, Socialism*, 258.

56. Poulantzas, *State, Power, Socialism*, 258.

57. Baiocchi, *We, the Sovereign.*

58. Bourdieu, *Firing Back.*

59. Gallucci, "Energy Equity."

60. Gallucci, "Energy Equity."

61. Angel, "Towards an Energy Politics." See also Baiocchi, *We, the Sovereign.*

62. Muñoz, *Disidentifications.*

63. Muñoz, *Disidentifications*, 11.

64. Koeppel, "Organizing for Energy Democracy."

65. Lieberman, "How This Man Is Helping Native Americans."

66. The US military is the single largest oil-consuming institution on the planet. See Union of Concerned Scientists, "The US Military and Oil."

67. Lefebvre and Adragna, "Trump Administration Seeks Criminal Crackdown."

68. Roberts, "New York Just Passed."

69. Con Edison, "Renewable Energy Systems."

70. For some typical arguments in favor of horizontalism, see Sitrin, *Horizontalism*.

71. Grimley, "Just How Democratic."

72. Wells, interview by author.

73. One Voice, "Electric Cooperative Leadership Institute."

74. Matt Grimley, "Just How Democratic."

75. Kauffman, "She Begged for Mercy."

76. Cohen, "We Have to Finance a Global Green New Deal."

77. LeMenager, *Living Oil*.

78. Yusoff, *A Billion Black Anthropocenes or None*.

References

Amazon Defense Coalition. "Chevron Suffers Major 8–0 Defeat in Ecuador's Constitutional Court Over Landmark Pollution Judgment." *CSR Wire*, https://s3.amazonaws.com/fcmd/documents/documents/000/004/827/original/ChevronTexaco_-_Aguinda_July_2018_update_ADCPR.pdf?1532349655.

Angel, James. "Towards an Energy Politics in-against-and-beyond the State: Berlin's Struggle for Energy Democracy." *Antipode* 49, no. 3 (2017): 557–76.

Asamblea Nacional. "Ecuadorian Constitution." https://www.asambleanacional.gob.ec/documentos/constitucion_de_bolsillo.pdf.

Baiocchi, Gianpaolo. *We, the Sovereign*. Medford, MA: Polity, 2018.

Bourdieu, Pierre. *Firing Back: Against the Tyranny of the Market*. New York: Verso, 2002.

Boyer, Dominic. "Infrastructure, Potential Energy, Revolution." In *The Promise of Infrastructure*, edited by Nikhil Anand, Akhil Gupta, and Hannah Appel, 223–44. Durham, NC: Duke University Press, 2018.

Breakthrough Institute. "An Ecomodernist Manifesto." Ecomodernism. https://www.ecomodernism.org (accessed February 24, 2021).

Bullard, Robert. "Environmental Justice in the 21st Century: Race Still Matters." *Phylon* 49, nos. 3–4 (2001): 151–71.

Caffentzis, George. "A Discourse on Prophetic Method: Oil Crises and Political Economy, Past and Future." In *Sparking a Worldwide Energy Revolution: Social Struggles in the Transition to a Post-Petrol World*, edited by Kolya Abramsky, 60–71. Oakland, CA: AK, 2010.

Checker, Melissa. "Wiped Out by the Greenwave: Environmental Gentrification and the Paradoxical Politics of Urban Sustainability." *City and Society* 23, no. 2 (2007): 210–29.

Cohen, Rachel. "We Have to Finance a Global Green New Deal—Or Face the Consequences." *Intercept*, July 24, 2019. https://theintercept.com/2019/06/24/global-green-new-deal-climate-finance/.

Con Edison. "Renewable Energy Systems" https://www.coned.com/en/our-energy-future/renewable-energy-systems.

Coryat, Diana. "Extractive Politics, Media Power, and New Waves of Resistance against Oil Drilling in the Ecuadorian Amazon: The Case of Yasunidos." *International Journal of Communication* 9 (2015): 3741–60.

Dawson, Ashley. *Extreme Cities: The Perils and Promise of Urban Life in the Age of Climate Change*. New York: Verso, 2017.

Fairchild, Denise, and Al Weinrub. Introduction to *Energy Democracy: Advancing Equity in Clean Energy Solutions*, edited by Denise Fairchild and Al Weinrub, 1–20. Washington, DC: Island, 2017.

Gallucci, Maria. "Energy Equity: Bringing Solar Power to Low-Income Communities." *Yale Environment 360*, April 4, 2019.

Gilmore, Ruth Wilson. *Golden Gulag: Prisons, Surplus, Crisis, and Opposition in Globalizing California*. Berkeley: University of California Press, 2007.

Gómez-Barris, Macarena. *Beyond the Pink Tide: Artistic and Political Undercurrents*. Oakland: University of California Press, 2018.

Gómez-Barris, Macarena. "The Colonial Anthropocene: Damage, Remapping, and Resurgent Resources." *Antipode Online*, March 19, 2019. https://antipodeonline.org/2019/03/19/the-colonial-anthropocene/.

Gómez-Barris, Macarena. *The Extractive Zone: Social Ecologies and Decolonial Perspectives*. Durham, NC: Duke University Press, 2017.

Greenwald, Glenn. "Watch: Glenn Greenwald's Exclusive Interview with Bolivia's Evo Morales, Who Was Deposed in a Coup." *Intercept*, December 16, 2019. https://theintercept.com/2019/12/16/evo-morales-interview-glenn-greenwald/.

Grimley, Matt. "Just How Democratic Are Rural Electric Cooperatives?" Institute for Local Self-Reliance, January 13, 2016. https://ilsr.org/just-how-democratic-are-rural-electric-cooperatives/.

Harasim, Kinga. "Bolivia's Lithium Coup." *Latin America Bureau* (blog), December 8, 2020. https://lab.org.uk/bolivias-lithium-coup/.

Harvey, David. *The Conditions of Postmodernity: An Enquiry into the Origins of Social Change*. New York: Wiley-Blackwell, 1991.

Hickel, Jason. "The Myth of America's Green Growth." *Foreign Policy* (blog), June 18, 2020. https://foreignpolicy.com/2020/06/18/more-from-less-green-growth-environment-gdp/.

Holmes, Colin. "The Cost of Utilities: Which States Pay the Most?" Move.org, January 29, 2018. https://www.move.org/which-states-pay-most-utilities/.

Jameson, Fredric. *Postmodernism, or The Cultural Logic of Late Capitalism*. Durham, NC: Duke University Press, 1991.

Kauffman, Alexander. "She Begged for Mercy: The Utility Cut Her Elderly Parents' Power Anyway." *HuffPost*, April 14, 2020. https://www.huffpost.com/entry/coronavirus-utilities-north-carolina_n_5e836400c5b6d38d98a589b7.

Koeppel, Jackson. "Organizing for Energy Democracy in the Face of Austerity." Transnational Institute, April 12, 2019. https://www.commondreams.org/views/2019/04/14/organizing-energy-democracy-face-austerity.

Lefebvre, Ben, and Anthony Adragna. "Trump Administration Seeks Criminal Crackdown on Pipeline Protests." *Politico*, May 3, 2019. https://www.politico.com/story/2019/06/03/trump-administration-seeks-criminal-crackdown-on-pipeline-protests-1499008.

LeMenager, Stephanie. *Living Oil: Petroleum Culture in the American Century*. New York: Oxford University Press, 2014.

Lennon, Myles. "Decolonizing Energy: Black Lives Matter and Technoscientific Expertise Amid Solar Transitions." *Energy Research and Social Science* 30 (2017): 18–27.

Lieberman, Bruce. "How This Man Is Helping Native Americans Gain Energy Independence." *Yale Climate Connections*, June 7, 2018 https://yaleclimateconnections.org/2018/06/native-americans-move-toward-energy-independence/.

Martinez, Esperanza. "Yasuní El tortuoso camino de Kioto a Quito." Quito: CEP y Abya-Yala, 2009.

McAfee, Andrew. *More from Less: The Surprising Story of How We Learned to Prosper Using Fewer Resources—And What Happens Next.* New York: Simon and Schuster, 2019.

McMullen-Laird, Lydia. "End of 'Net Metering' Casts Shadow over Future of New York Solar." *Indypendent*, September 4, 2018.

Mignolo, Walter. "Delinking: The Rhetoric of Modernity, the Logic of Coloniality, and the Grammar of De-coloniality." *Cultural Studies* 21, nos. 2–3 (2017): 440–514.

Mitchell, Timothy. *Carbon Democracy: Political Power in the Age of Oil.* New York: Verso, 2011.

Muñoz, José Esteban. *Disidentifications: Queers of Color and the Performance of Politics.* Minneapolis: University of Minnesota Press, 1999.

Odum, Howard T. *Environment, Power, and Society.* New York: John Wiley, 1971.

One Voice. "Electric Cooperative Leadership Institute." https://energydemocracy.us/one-voice/.

Oveido Freire, Atawallpa. *Buen vivir (Sumak Kawsay): "Tercer Estado," Alterativa al capitalismo y al aocialismo.* 5th ed. Independently published, 2019.

Pérez-Medina, Lourdes, and Elizabeth Yeampierre. "The People's Power." *Urban Omnibus*, April 10, 2019. https://urbanomnibus.net/2019/04/the-peoples-power/.

Pitron, Guillaume. *The Rare Metals War: The Dark Side of Clean Energy and Digital Technologies.* Minneapolis: Scribe, 2020.

Political Database of the Americas. "Constitution of the Republic of Ecuador." 2008. Last updated, January 31, 2011. https://pdba.georgetown.edu/Constitutions/Ecuador/english08.html.

Poulantzas, Nicos. *State, Power, Socialism.* New York: Verso, 1978.

Quijano, Anibal. "Coloniality of Power, Eurocentrism, and Latin America." *Nepantla: Views from South* 1, no. 3 (2000): 533–80.

"Rights of Nature Articles in Ecuador's Constitution." Title II. https://therightsofnature.org/wp-content/uploads/pdfs/Rights-for-Nature-Articles-in-Ecuadors-Constitution.pdf.

Roberts, David. "New York Just Passed the Most Ambitious Climate Target in the Country." *Vox*, July 22, 2019. https://www.vox.com/energy-and-environment/2019/6/20/18691058/new-york-green-new-deal-climate-change-cuomo.

Ruivenkamp, Guido, and Andy Hilton. Introduction to *Perspectives on Commoning: Autonomist Principles and Practices*, edited by Guido Ruivenkamp and Andy Hilton, 1–21. London: Zed Books, 2017.

Sacher, William. Interview with Macarena Gómez-Barris, May 18, 2015.

Salminen, Antti, and Tere Vadén. *Energy and Experience: An Essay in Nafthology.* Chicago: MCMPrime, 2015.

Scheer, Herman. *The Solar Economy: Renewable Energy for a Sustainable Global Future.* London: Earthscan, 2004.

Secretaría Nacional de Planificación y Desarrollo. *Buen Vivir: 2017–2021.* https://www.gobiernoelectronico.gob.ec/wp-content/uploads/downloads/2017/09/Plan-Nacional-para-el-Buen-Vivir-2017-2021.pdf.

Sitrin, Marina. *Horizontalism: Voices of Popular Power in Argentina.* San Francisco: AK, 2006.

Smil, Vaclav. *Energy in Nature and Society: General Energetics of Complex Systems.* Cambridge, MA: MIT Press, 2008.

Stein, Sam. *Capital City: Gentrification and the Real Estate State.* New York: Verso, 2019.

Tainter, Joseph A. *The Collapse of Complex Societies*. Cambridge: Cambridge University Press, 1988.

Telesur. "Elon Musk Confesses to Lithium Coup in Bolivia." July 25, 2020. https://www.telesurenglish.net//news/elon-musk-confesses-to-lithium-coup-in-bolivia-20200725-0010.html.

Union of Concerned Scientists. "The US Military and Oil." June 30, 2014. https://www.ucsusa.org/resources/us-military-and-oil.

United Nations. "SDG Indicators." https://unstats.un.org/sdgs/report/2019/goal-12/ (accessed February 24, 2021).

Wallace, Mike. *A New Deal for New York*. New York: Bell and Weiland, 2002.

Wells, Benita. Interview by author, February 23, 2018.

Yaeger, Patricia. "Editor's Column: Literature in the Ages of Wood, Tallow, Coal, Whale Oil, Gasoline, Atomic Power, and Other Energy Sources." *PMLA* 126, no. 2 (2011): 305–26.

YASunidos International. "YASunidos Manifesto." https://yasunidosinternational.wordpress.com/manifiesto/.

Yusoff, Kathryn. *A Billion Black Anthropocenes or None*. Minneapolis: University of Minnesota Press, 2019.

From Waste to Climate

Tackling Climate Change in a Rebel City

Marco Armiero

Naples: Wasted City

From 1994 to 2012, Naples and its region were under a special administrative and legal regime due to a so-called waste emergency.[1] Everything started from the fact that most of the landfills of the region were exhausted or working under illegal and unsanitary conditions. For this reason, most of them were closed by the judiciary, leaving the metropolitan city of Naples without a functioning system for the disposal of waste. Almost immediately, the streets of Naples started to be flooded by urban garbage; heaps of trash were piling up, sometimes reaching the first floors of the buildings. Journalists from every corner of the world arrived in Naples, attracted by the lure of this urban apocalypse. Mothers zigzagging with their children among piles of garbage became an iconic image of the city.[2] When the trash reached the posh areas of Naples, waste became visible, palpable enough to be televised. Nonetheless, the tension between what was visible and what remained hidden stayed unsolved. The underground flows of toxicity, the unjust distribution of burdens, and the effects of all this on human health needed a different way of seeing and understanding the socioecological relationships producing not the garbage in the streets but the wasting of subaltern people and communities. The combination Naples/waste is a staple in the global imaginary about the city. Almost all foreign travelers visiting Naples in the nineteenth century would remark on the dirtiness of the streets, the cohabitation of humans and garbage, often making a disturbing connection between this external disorder and some kind of innate characteristics of Neapolitans. In this respect, the narratives about Naples seem to follow the typical colonial/modernist trope described by Dipesh Chakrabarty for Indian cities:[3] dirtiness and

Social Text 150 · Vol. 40, No. 1 · March 2022
DOI 10.1215/01642472-9495117

diseases go hand in hand with the local resistance to the modern order that separates the dirty and the clean, the private and the public, the healthy and the sick. The same kind of Orientalist discourse can be found in Alfred Sohn-Rethel's 1926 description of the Neapolitans as premodern subjects in their almost superstitious approach to technology.[4]

Later on, several times during the nineteenth and twentieth century, the love story between Naples and its waste has become more visible, producing minor or impressive apocalypses. In 1884 a cholera epidemic hit the city, unleashing an impressive number of social scientists, all roaming around the ruins of a never fully modern metropolis. Whatever the etiology of the epidemic was, it was clear that the dirtiness of the city, the disordered blend of any kind of matter, the cohabitation of waste and people were crucial for the development of contamination.

In 1973, the cholera epidemic again visited Naples, bringing with it the usual parade of reporters and writers, ready to resuscitate the waste pornography of a city soaked in its own filth. Hence, the 1990s–2000s waste emergency was not an unexpected apocalypse but rather an epiphany confirming what was already known: Naples has always been a dirty city where waste overflows everywhere, contaminating places and bodies. In the descriptions of urban environmental disasters, the dirtiness of a place is almost always connected to the dirtiness of its inhabitants. After all, the Italian historian Benedetto Croce—himself a Neapolitan by adoption—crystallized this vision about the city with his famous definition of Naples as "a paradise inhabited by devils."[5] Croce meant to contrast the astonishing beauty of the city with what he considered its major problem—that is, the completely unruly, almost wild character of its inhabitants. The stigma against Neapolitans, and more broadly Southern Italians, has a long history, and found its apex with the unification of the country in the 1860s, when several scholars tried to demonstrate the inferiority of people from the Italian South on a "scientific" basis.[6] Things have not changed so much since then, considering that during the waste emergency many commentators and politicians went back to the reassuring narrative of the paradise inhabited by devils. Even scholars are not immune from this. Once a reviewer of one of my publications did not have any problem writing that I did not consider in my text the reluctance of Neapolitans to recycle and their habit of littering. Neapolitans are dirty, and an epidemic or a waste emergency are clear exemplars of their "nature"—this seems to be the mainstream narrative about the city.

Instead of these trite stereotypes, one might read these various emergencies as demonstrations not of some natural disposition of the city inhabitants but of long-lasting structures of injustice creating vulnerable people and wasted places. The entire body of scholarship and activism on environmental justice has argued precisely that if people live in unhealthy environments, most of the time it is due to the unequal power relationships that have targeted them as the ultimate dump for someone else's well-being. While it is crucial to restore the truth, I argue that it is still insufficient to frame these waste crises only in terms of victims, subaltern people forced to endure contamination as the price to pay for capitalism's organization of space and production. Waste crises are also the forcing house of powerful community infrastructures, catalytic moments of massive social mobilizations.[7] In my research on the Neapolitan case, I have argued that the embodied experience of contamination has caused a political subjectification creating resistant communities and activist knowledge. Since the waste emergency exploded, several grassroots organizations have mushroomed both in the metropolitan region of Naples and in the entire region. Most of these grassroots organizations were connected to specific sites, mobilizing to stop the construction or enlargement of waste facilities in their communities. This might seem a typical NIMBY (not in my backyard) approach to environmental problems, and indeed this is how those mobilizations have been depicted by mainstream media. After all, hordes of wild Neapolitans rioting against "rational" and "modern" waste infrastructures fit perfectly well with the age-old adage about a paradise inhabited by devils with a complicated relationship with technology, as Alfred Sohn-Rethel would add. It matters little that those infrastructures—from incinerators to landfills—were much less modern and rational than it was claimed or that they were always placed in communities that had already paid a high price to keep other parts of the region clean.

In this article, I wish to explain how this waste struggle legacy has intertwined with the current climate activism. Both waste and climate are crucial ecological problems; the exponential increase in CO_2 emissions could be considered in itself a problem of disposal of waste.[8] But in many ways waste and climate are radically different ecological problems; issues of scales—both geographical and temporal—and perceptions and scientific frameworks affect the kind of social activism they may generate. Rather than exploring the relationships between mainstream and grassroots environmentalism, something that has been analyzed so many times in the literature,[9] I aim at interrogating what may occur when a community with a long history of environmental and social justice struggles meets climate change. And I will do so from a personal, situated point of view, since I myself am involved in environmental and political activism in Naples, Italy.

The Performative Politics of Waste

In my attempt to understand the continuity between waste activism and climate activism, I believe that there are two notions that are crucial: the construction of epistemic communities and the rejection of single-issue politics. According to urban theorist Edgar Pieterse, the production of knowledge, and thereby of new imaginaries, occurs through "epistemic communities" where collective knowledge "can be assembled or networked" with the aim to "challenge conventional orthodoxy (the mainstream) about what is possible and not possible in terms of transformative urban development agendas."[10] I argue that such epistemic communities have been the most remarkable result of the long-lasting waste struggles in Naples. Because while the mainstream pornography of waste generates an easy understanding of what we are talking about—the heaps of trash and the smell in the air—social struggles bring more nuances into the picture. As I have argued elsewhere, looking at environmental issues through the lens of conflicts "can reveal the structure of power embodied in nature as well as the socially diversified contents of humans' agency."[11] The geography of landfills, incinerators, and toxic waste does not only speak of technological solutions and environmental contamination; it also makes visible the hierarchies of power relationships between groups and places. Looking at waste through socioecological struggles, it is clear that waste is never just a matter of trash. Every environmental justice conflict is at its core a scientific controversy. In the Neapolitan case, activists had to become experts of urban and industrial ecologies in a spasmodic attempt to restore the truth and connect the dots linking health, contamination, production, and corruption.[12] Accused of being antiscience and primitive for their critique of incinerators and their skepticism toward mainstream scientific knowledge (not so different from the charges leveled in 1926 by Sohn-Rethel), activists were rejecting neither technology nor science. Rather they tried to appropriate and critically analyze that knowledge while aiming to coproduce new knowledge for radical change. Often this has led to the uncovering and mobilization of the plurality of science, to discovery and even fostering of radical divergences between scientists—indeed, on several themes not all scientists say the same thing. In many cases, activists have been instrumental in the production of militant knowledge, for instance through the experiences of popular epidemiology or citizens' monitoring of air quality. On a more epistemological level, activists have worked to question the usual binary opposition between impersonal scientific knowledge and embodied personal experiences, claiming that the two are more deeply connected than is usually acknowledged.[13] The activists' crucial contribution to a more accurate understanding of the waste crisis in Campania brought a shift from focusing on bags of urban trash in the streets to highlighting the illegal disposal of toxic waste in the rural areas around the metropolis. In other words, although with climate the challenge seems to make visible an invisible threat, with waste, activists needed to shift the attention from the cumbersome presence of waste in the city streets to the subtle infiltration of toxicity into bodies and ecosystems. It was still an issue of visibility/invisibility, as is often the case with environmental issues, but I argue that the waste struggles produced an epistemic community trained to navigate between the two, equipped to look for underlining power relationships within ecological problems. With Venn, I would argue that the waste struggles provided activists with an embodied theory of waste that brought "into visibility the constitutive relation between the visible and the invisible."[14]

In 2006 the intellectual fulcrum of the Neapolitan epistemic communities, the Assise,[15] published a pamphlet, with the unequivocal title *Warning: Toxic Waste* (original title: *Allarme Rifiuti Tossici*), in which the activists aimed to switch public attention from the extremely visible mess caused by urban garbage to the invisible though more harmful flow of toxic substances.[16] As Rob Nixon has argued, slow violence goes hand in hand with its invisibility, or, to be more precise, with its "invisibilization."[17] Restoring the truth has been the main mission of the Neapolitan epistemic communities. While mainstream media and politicians were deriding those who were protesting and denying any connection between their health conditions and the disposal of toxic waste, epistemic communities were at work making research and gathering materials that could prove their claims. Every Sunday for seven years (2005–11), the Assise provided a venue where the Neapolitan activists could meet with sympathetic experts (medical doctors, geologists, jurists, and others) for a self-managed education. It is enough to read the documents produced by the two wider coalitions of civil society organizations (the Waste Regional Coordination and the Stop Biocidio Network) to have a sense of the knowledge produced by these epistemic communities.

Comparing these two big coalitions can help to highlight the other point I believe is crucial for understanding the connections between waste struggles and climate activism—that is, the rejection of single-issue politics. Green parties and environmentalist organizations have often been defined as handbook examples of single-issue politics; they are supposed to target ecological problems and nothing else. The entire environmental justice movement started from a radical critique of the mainstream environmentalist organizations, accused of being blind to—if not complicit with—structural inequalities related to class, race, and gender. I argue that while the Waste Regional Coordination has been true to its mission, focusing exclusively on the issue of waste, the Stop Biocidio Network has chosen to stay in the Italian radical leftist tradition by merging environmental and social issues. In the end, the former has now stopped its activities, while Stop Biocidio still is the most lively coalition of environmentalist grassroots associations in the region and perhaps on the national scale, playing a crucial role both in the struggles for the reclamation of toxic communities and in the new mobilization for climate justice. In a document written in preparation for a national meeting held in Naples, Stop Biocidio activists wrote:

> Capitalism is the virus infecting our communities and the biocide is the disease brought by this virus. Biocide is the pathological development coming from capitalism because capitalism is in direct contradiction to life. Climate change is the most transversal symptom of this disease. We are not all equally responsible for climate change, we are not all victims and guilty parties in the same way, as if the problem were the human species. Instead there are a few who have enriched themselves exploiting both the resources of the planet and the majority of people.[18]

Biocide as the theoretical category to understand the socioecological crisis turns out to be a key bridge between waste and climate. Thinking with biocide implies attributing responsibilities; biocide does not just occur but it is implemented. The term *biocide* helps avoid the usual identification between symptoms and causes, or, in other words, the reification of socioecological problems as if waste or CO_2 emissions were what should be solved. While biocide brings power structures to the core of the ecological thinking, it also opens up to a more-than-human understanding of the crisis. Although generated from the experience of human sufferings, biocide frames that experience in terms of a broader violence exerted by capitalism against life. Indeed, after years of waste struggles, the Stop Biocidio coalition was not only organizationally prepared for entering into the climate movement; the category of biocide gave it the intellectual tools for bridging local antitoxic campaigns and global climate actions. Something similar occurred in New Orleans, when environmental justice organizations started to engage with climate change after Hurricane Katrina, bringing into that arena their long-lasting commitment to social justice.[19]

By hosting the national meeting of March 3, 2019, in Naples, Stop Biocidio situated itself as the hinge between environmental justice struggles in the city and the growing mobilization against climate change. Nonetheless, it was not an immediate, easy transition; employing the category of biocide to understand and act on climate change needed some time and thinking.

The Two Tales of One Organization

When I started my research on climate change activism in Naples in 2018, it was an obvious choice for me to begin from the groups I had worked with since the time of my research on waste. I convened a focus group with the youth branch of a grassroots organization I became close with during my many years of fieldwork. Around twelve students—mostly high

Figure 1. Stop Biocidio banner at the Climate Strike in Naples, September 27, 2019. Photograph by the author.

school students—came to the meeting in a squatted room at the local university. They knew me quite well since I have offered them several classes for self-education on political ecology and environmental justice; however, when we met this time they did not want to be recorded, and I could perceive they were not so happy about the theme of the meeting. As they clearly stated, climate change was not their priority; it was something distant from their lives and their struggles. One of them, A., explained that climate change was something she had learned about at school, where the science teacher told her about the melting ice of the poles, while no teacher ever discussed, at school, the issue of toxic contamination in the region. For this activist, this was a clear sign that climate change was a conflict-free zone, an area she was not interested in being engaged with. Indeed, these students were not the usual environmental activists; rather they were militants of a radical leftist organization, with quite strong ties to Italian autonomia and inspired by the Zapatista and Kurdish insurgencies. Nonetheless, during the 2000s their organization became a key player in struggles for environmental justice in Naples, mainly because their headquarters—an abandoned primary school they have occupied illegally since 2003—is located in the working-class neighborhood Chiaiano, where a gigantic landfill was planned and built. Theirs was a place-based struggle, though not at all a NIMBY one, since they were behind several coalitions coordinating the struggles and mobilizations in various communities, starting with the Rete Commons (Commons Network) and culminating with the still-active Stop Biocidio. Hence, the objection of A. was perfectly in line with an activist tradition and culture that privileges place-based struggles and an antagonist agenda. At that time, these activists were also supporting the struggles against the Trans-Adriatic Pipeline (TAP)[20] occurring in the neighboring region of Apulia—proving once more that theirs was not an egoistic/NIMBY approach. However, when I asked them why they were not framing that struggle in terms of climate change, they argued that it was much better to think of it in terms of territorial struggles. The TAP was jeopardizing the local ecosystem, with its centuries-old olive trees, amazing beaches, and crystal sea, and those were the main issues raised by local activists. Being in solidarity with them, the Neapolitan activists did not consider it appropriate to bring themes to the table that were not locally developed. On a more general level, my interlocutors argued that climate change was an excessively abstract theme that people would not feel particularly connected to. In addition to this, they also raised the issue of the absolute disproportion between the possible actions done at the grassroots level and the magnitude of the problem. The situation was even more depressing since they have been part of a progressive coalition that had won the municipal elections twice, ruling the city since 2013.[21] For them the local government could not do anything to address climate change.

I left Naples with the impression that radical environmentalist organizations (REOs) were not interested in the climate struggle. As much as they had become acquainted with ecology during the garbage struggles of the early 2000s, they did not seem willing to embrace what appeared to them a global issue, too ecumenical to become terrain for social conflicts. The situation was different for mainstream environmentalist organizations (MEOs), such as the World Wildlife Fund and the Italian NGO Legambiente. In the focus groups I held with them, they seemed more prepared to embrace climate change as a major theme for their actions. They

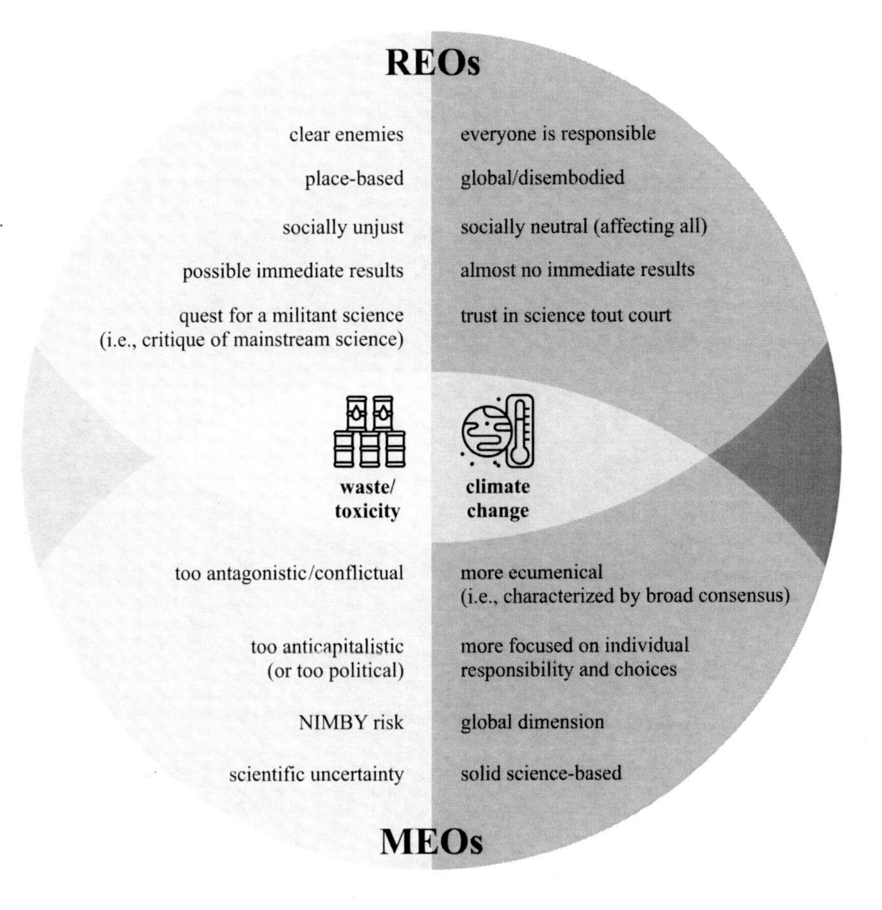

Figure 2. Radical Environmental Organizations (REOs) vs. Mainstream Environmental Organizations (MEOs)—Dealing with climate change and waste/toxicity. Graphic design by Simona Quagliano.

were involved in national and international campaigns on climate change, and they were also becoming engaged in local campaigns to promote the greening of the city, now reframed in terms of mitigation strategies. The divergence between these two kinds of environmentalist organizations did not surprise me at all. Having worked for a long time on the histories of environmentalism, I am familiar with the "varieties of environmentalism" that have characterized the evolution of the movements.[22] But it was still a revelatory experience to notice that there was some kind of inverse correlation between the rate of involvement in local struggles and the willingness to embrace climate change as a major theme for mobilization. In fact, mainstream environmentalist organizations such as WWF and Legambiente were almost absent in the cycle of struggles over waste and toxic contamination, at least at the grassroots level. My impression was that the main problems felt by the radical activists in respect to climate

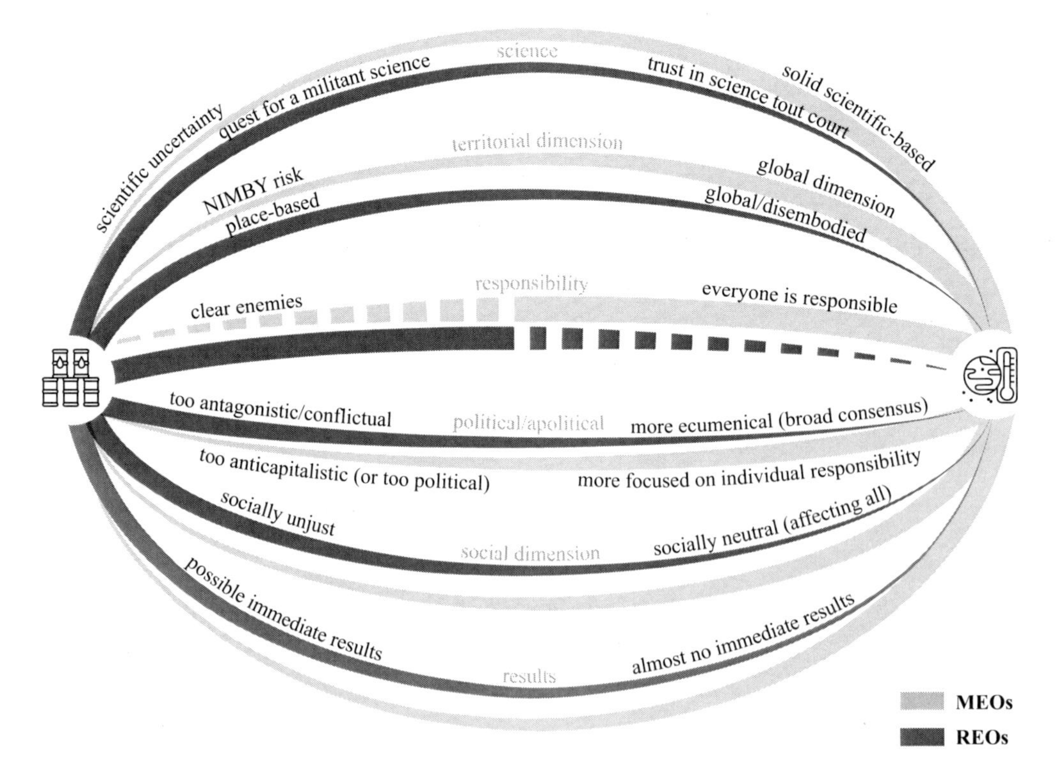

Figure 3. Engaging with toxicity and climate change—REOs and MEOs. The intensity of the engagement is represented by the thickness of the lines; the thicker the line, the more engaged the organization. Graphic design by Simona Quagliano.

change were precisely the main reasons for those mainstream organizations to tackle it.

The REOs made an excoriating critique of the established arsenal mobilized in climate change debates and activism: the individualization of responsibility, the neutrality of its effects, and the search for technological solutions. I did not ask the MEOs to elaborate on their reluctance to fully engage with the waste crisis but, building on more than a decade of research on the theme, I would imagine that their answer would have mirrored this: it had become too divisive, and it was monopolized by radical activists with a repertoire of actions that were too far removed from their usual practices. Instead, MEOs were much more comfortable in the climate change arena, where they seemed to have the right tools and approaches to tackle the crisis. Another and perhaps key point of divergence is the relationship with science and scientists, since climate change activism—at least its mainstream version—relies heavily on scientific knowledge to frame its claims as nonpartisan, while the antitoxic activists often have to challenge mainstream scientists, who are generally reluctant

Figure 4. Life vs. Capital—training school for climate activists. Photograph by the author.

to acknowledge their claims. The divergence between the two strands of environmentalism could not be clearer in front of my eyes. But, in the following months, everything changed.

In March 2019 the same association I had met with in the fall hosted a national assembly in Naples with the aim of connecting place-based environmental struggles to the growing climate change movement. The assembly was hosted in the city hall with the discreet but clear support of the municipal government. Since then, it has become a key player in the Neapolitan and national Fridays for Future (FFF) movement. The weekly meetings of the city chapter have been held in its office, and it was the only organization from Naples to participate in the Venice Climate Camp in 2019. Finally, it was instrumental in the organization of a climate change training camp in Naples, again in the fall of 2019, which preceded the national meeting of FFF Italy, held in Naples on October 6, 2019.

Changing the Climate of the Movement

I wonder what happened in the few months between my first focus group in the fall of 2018 and the new trajectory that places this REO at the very core of the climate change mobilization.

An easy answer to the question could be that what happened was

Greta Thunberg. Starting from 2018, the Swedish climate activist radically changed the landscape of environmental activism. Some observers have spoken of a "Greta effect." When she started to protest every Friday in front of the Swedish parliament, sitting alone with a poster in her hands, I thought she was the best example of the Nordic version of activism, one based on individual choices not only in terms of consumerism but even of social mobilization. The difference with the environmental movements I have studied in Italy and also in other southern European countries was remarkable. In my research in Italy, I have never encountered this kind of individual mobilization but rather the collective construction of resisting communities. However, in the case of Greta, what had started as an individual protest relatively quickly became a global mass movement, Fridays for Future.

The Greta effect, though, can explain only partially the radical change experienced by a grassroots organization as Stop Biocidio. I have also detected some skepticism or even open hostility toward not so much Greta, per sè—that kind of hostility has been more common among the extreme right—but against the type of environmentalism that she embodies. In a debate hosted by the Italian leftist online magazine *La sinistra quotidiana* (*The Daily Left*), all four of the articles were quite critical of Greta Thunberg's environmentalist agenda. Although with different degrees of criticism, the authors stressed the apolitical nature of Thunberg's environmentalism, which they accused of not being explicitly anticapitalist.[23] Other leftist intellectuals have voiced similar reservations, for instance the French Marxist philosopher Georges Gastaud, who blamed the apolitical nature of the movement generated by the Swedish activist.[24] However, it is true that Thunberg has also encountered positive reactions in the leftist circles; for instance, both Slavoj Žižek[25] and Naomi Klein[26] have expressed their sympathies for the young Swedish activist, while on a more institutional level Alexandra Ocasio-Cortez has also supported Thunberg's activism.[27]

In the Italian case, as always with social movements, there have been attempts to co-opt or domesticate the FFF movement. In 2019 the new secretary of the Democratic party (Partito Democratico in Italian) dedicated his election as the chairman of the party to Thunberg, thereby committing his leadership to the struggles against climate change. The day after his election, he also went to Turin to stress his support for the construction of the extremely controversial high-speed train between Turin and Lyon,[28] the blocking of which has become a key symbolic fight for the Italian REOs.[29] The risk of co-optation from the political establishment became even more evident during Thunberg's visit to Italy in April 2019. On that occasion she met the president of the senate and received the honorary membership of Italy's largest union organization; both moves were not welcomed positively by Italian activists, who feared supporting the greenwashing of these mainstream institutions.

Thunberg's presence in Rome—that is, at the core of Italian political power—revealed some frictions that are constitutive rather than episodic in the national climate change movement. And, I would argue, it may also explain the "conversion" of an REO such as Stop Biocidio; instead of merely observing the contradictions within the climate movement, it has decided to enter into the arena and foster its radical agenda. During my fieldwork in Naples, especially in the preparation for the training school Life vs. Capital, it became clear to me that FFF was traversed by diverse political—or self-declared apolitical—cultures. I was deeply involved in the organization of the training school, but in the process it became more and more evident that something was wrong. FFF official social channels did not include any news about the school, thereby avoiding endorsing it. It was not easy to decipher the jungle of signals that populated the communications among activists in the hectic preparation for the training school and the FFF national assembly. In the end, the school became the expression of only a few of the grassroots organizations that are part of the FFF network, mainly the most politically radical and explicitly connected to an anticapitalist agenda. Rumors arrived that in the FFF national coordination some were uncomfortable with the school's name: Life vs. Capital. This does not mean the school was a failure; more than one hundred youths participated in the courses, and not only well-trained activists but also students without any political commitment who were reached thanks to an agreement with a local high school.[30]

I believe that the Life vs. Capital training school became involved in a wider confrontation between diverse political tendencies present in the Italian FFF network, a confrontation that everybody knew would inevitably explode once it was decided to bring the FFF national assembly to Naples. In fact, Naples is without a doubt the emblem of Italian radical environmentalism; as I have illustrated above, twenty years of struggles over toxic contamination merged with a long tradition of leftist activism have created a radical environmental culture that has even produced an institutional result with the election of a radical leftist local government.

In this sense, the history of the Stop Biocidio coalition illustrates a dialectic relationship with what has been called the Greta effect. Activists in this coalition did not join a prepackaged movement but entered into the coalition with their own agenda and the ambition to affect its possible direction. The final document of the Neapolitan national assembly represented the efforts of the most radical groups to provide a political agenda for the movement.[31] The document affirms the unity of all the struggles against oppression and exploitation, defining FFF as an antifascist, antiracist, and trans-feminist movement, a move that seems not different from the intersectional alliance proposed, for instance, by Giovanna DiChiro in the name of social reproduction.[32] The document also supports the need for an alliance between environmentalists and workers rejecting the job blackmail that forces people to choose between health and salary. It is true that the word *capitalism* is not present in the final report of the assembly, and from my fieldwork I know that including an explicit anticapitalist agenda was at the center of an extremely heated debate inside the group working on that document.[33] The option for an explicit anticapitalist politics was mentioned in several of the speeches and in the reports of some of the working groups, including the one on collective practices:

> Although the movement is split on the issue of embracing an anti-capitalist agenda, it is true that Greta herself has pronounced an excoriating critique of capitalism in her speech at the United Nations. . . . *We do not want to use anti-capitalism just as an empty* identitarian label, rather we want to express our anti-capitalism in our practices and direct actions against extractivist capitalism.[34]

Tellingly, a few lines below, the same document stated that the struggles to save humans from extinction should be connected to the struggles for a decent life. Finally, the activists gathered in Naples wrote that individual actions are useful but not sufficient to stop climate change, committing FFF to fight against the capitalist organization of society.

This stress on practice was partially a device to bring an anticapitalist agenda into the FFF debate, avoiding the frictions caused by that kind of politics among the participants in the national assembly. In other words, the reports from the working groups—and one was dedicated precisely to practices—did not need to be approved by the general assembly; thereby, they were easily steered toward more radical positions. However, I would argue that this was not only a tactical device to include some otherwise controversial slogans. Rather, I believe that the choice to include the anticapitalist agenda in a document on the movement's practices represents a strategic posture. Some REOs, and perhaps especially Stop Biocidio, have always been skeptical of what they call "identitarian politics," meaning not the identity politics of the New Left but rather an attachment to political minoritarian identities, which can be quite strong among Italian leftist organizations. In this sense, *identitarian politics* means a reproduction of old symbols and slogans, without any attempt to reinvent them in tune with the challenges of the present. In opposition to that approach, REOs like Stop Biocidio have interpreted radicalism more in their practices than in what they have often defined as "liturgies from the past." Instead of labels and slogans, they proposed radical political practices; for instance, during the national assembly in Naples the local organizers, mostly activists from Stop Biocidio and other REOs, organized a blockade of the Q8 oil deposit, while during the Venice Climate Camp, also promoted by the most radical sectors of FFF, the activists decided to occupy the red carpet of the Venezia Film Festival.

Synergistic Compost Rather than Hegemonic Plans: A Conclusion

The friction between what I have called REOs and MEOs is neither new nor exclusive to the Neapolitan case. During the climate camp in Venice, an activist from the Susa Valley reported a heated debate over the opportunity to have the NO TAV flags at the FFF march in Turin. For parts of the climate movement, those flags—and the twenty-year struggle they symbolized—were controversial because they referred to a quite antagonistic, even violent, fight. Someone would even say that opposing the construction of a high-speed train has been detrimental to the cause of limiting CO_2 emissions. At the same climate camp another activist criticized the part of my speech in which I campaigned for building an alliance between workers and the climate movement, arguing that workers in polluting industries were part of the problem and not victims of the capitalistic organization of production, as I stated. Finally, during the occupation of the red carpet, a group of activists complained about some slogans we were singing against fascism and xenophobia, saying that we were driving attention away from climate change.

In this article, I have explored the history of the encounter between an REO such as Stop Biocidio and the climate change movement. I have illustrated how the intellectual and organizational know-how accumulated during almost twenty years of struggles against toxicity has been mobilized in the climate change movement. In particular, I argue that the category of biocide was instrumental in this transition. As an interpretative tool, biocide frames ecological problems in terms of symptoms and etiologies while rejecting the idea that we are all guilty in the same way. According to such a radical approach, toxicity and climate change are not the causes of socioenvironmental problems but are symptoms of the capitalist virus that accumulates profits by sorting out what/who has value and what/who can be disposed of in the name of profit. However, the attempt to shape the political agenda of the climate change movement—in particular of the FFF—does not reproduce the usual, demeaning practice of avant-garde hegemony. Stop Biocidio has entered the FFF movement with its political agenda and its assets of intellectual categories and bricolaged practices but also with a clear awareness that those are not enough to build the mass movement the climate crisis needs.

Thinking with the category of biocide, activists in Stop Biocidio were primed to leave their political comfort zone, engaging with the notion of "life" while acknowledging that capitalism does not oppress only subaltern humans. The encounter between Stop Biocidio—and other REOs in general—and the climate movement not only brought a more radical political agenda to the latter; it also changed the former.

The case of the Venice Climate Camp exemplifies this dialectic relationship between REOs and the climate movement. In organizing the Venice Climate Camp, the *centri sociali*[35] of the northeast decided to embrace values and practices that were not characteristic of their tradition, such as an all-vegan menu, a plastic-free camp, and water-saving practices for all the camp. Similarly, in Naples, the radical leftist organization that is the engine of the Stop Biocidio coalition has embraced a plastic-free philosophy for its festival, forcing all the participants to think about their own consumption practices. Perhaps the most remarkable sign of this mutual contamination between REOs and the climate movement occurred on October 18, 2020, when a coalition of *centri sociali* and antispeciest activists attacked a pen where several bears were held in captivity because of their behavior—getting too close to humans or their animals. The platform for that march clearly shows the mutual learning between a radical anticapitalist agenda and a more-than-human understanding of the ecological crisis:

> Just as human beings have always resisted oppression and discrimination, also all non-human animals do not tolerate imprisonment and exploitation, they attack to defend themselves and try to escape, sometimes with success. On the side of the rebel bears, this is the slogan that accompanies the mobilization . . . to ask for the release of the imprisoned bears: M49-Papillon, M57 and Dj3.[36]

I argue that these examples prove that rather than a one-way hegemonic project, we have witnessed a dialectic relationship in which a new hybrid, plural, and nonsectarian movement was born.

Marco Armiero is research director at the Institute for Studies on the Mediterranean, CNR (Italian National Research Council) and director of the Environmental Humanities Laboratory, KTH–Royal Institute of Technology, Sweden. He is the author of the book *Wasteocene: Stories from the Global Dump* (2021).

Notes

I acknowledge the Occupy Climate Change! project funded by FORMAS (Swedish Research Council for Sustainable Development) under the National Research Programme on Climate (contract 2017-01962_3). Many thanks to all the activists and colleagues who have welcomed me in Naples.

1. On the waste crisis in Campania see Armiero, "Garbage under the Volcano"; Armiero and D'Alisa, "Right of Resistance"; Iovino, "Naples 2008, or, the Waste Land."

2. Searching for images of Naples and waste in Google provides abundant proof of my argument; see, among many others: Smoltczyk, Ulrich, and Wassermann, "How the Mafia Helped Send Italy's Trash to Germany"; Nadeau "Naples Suffers as Garbage Piles Up"; Viggiano and Hornby, "President Piles Pressure on Italy Government over Naples Trash"; Fisher, "In Mire of Politics and the Mafia, Garbage Reigns."

3. Chakrabarty, "Of Garbage, Modernity, and the Citizen's Gaze."

4. Sohn-Rethel, "Das Ideal des Kaputten."

5. Croce, *Un paradiso abitato da diavoli.*

6. On this, see Verdicchio, "The Preclusion of Postcolonial Discourse"; D'Agostino, "Craniums, Criminals, and the 'Cursed Race'"; Armiero, *A Rugged Nation*, 62–75.

7. Armiero and De Angelis, "Anthropocene"; Armiero and Sgueglia, "Wasted Spaces, Resisting People."

8. See, for instance, Venn, "Rubbish, the Remnant, Etcetera," 44.

9. On these themes I can once again refer to the foundational work Martínez Alier and Guha, *Varieties of Environmentalism.*

10. Pieterse, *City Futures*, 149.

11. Armiero, "Seeing Like a Protester," 59.

12. Armiero, "Is There an Indigenous Knowledge in the Global North?"

13. In another article, I have analyzed the politicization of the official medical records gathered by the activists living in a heavily contaminated neighborhood in Naples: "Evidently, those medical records speak the language of the official science, and they even have the stamps and signatures of scientific knowledge—hospitals, laboratories, head physicians—nevertheless, they also tell personal stories, by definition private, even intimate, in their nature" ("Is There an Indigenous Knowledge," 7).

14. Venn, "Rubbish, the Remnant, Etcetera," 45.

15. The Assise della Città di Napoli e del Mezzogiorno (the Permanent Assembly of the City of Naples and Southern Italy) was the most influential and active group of professionals and academics engaged in the struggles over garbage in Naples. It served as a sort of self-education arena for the activists, holding a weekly meeting every Sunday and publishing a bulletin and several pamphlets. On the Assise, see Capone, "The Assemblies of the City of Naples."

16. Capone, Cuccurullo, and Micillo, *Allarme rifiuti tossici.*

17. Nixon, *Slow Violence.*

18. Stop Biocidio, "Piattaforma contro il biocidio."

19. Schlosberg and Collins, "From Environmental to Climate Justice," 361.

20. More information about this project and the local opposition in Cavallini "Trans Adriatic Pipeline (TAP) in Puglia, Italy."

21. This grassroots organization elected three members of the city council and one deputy mayor of a neighborhood, and since 2019 one of its leaders entered in the

municipal government as deputy for culture and tourism. With the 2021 municipal election, that radical left coalition dissolved and a very large center-left coalition won the election. The grassroots organization I am talking about was able to elect only one city councilor.

22. Martínez Alier and Guha, *Varieties of Environmentalism*; Armiero and Sedrez, *History of Environmentalism*.

23. Fagan, "In direzione ostinata e contraria"; Il simplicissimus, "Timeo Gretas et dona ferentes"; Bonora, "Greta e i . . . gretini;" Zhok, "Greta e il problema."

24. Redazione, "Si può criticare Greta Thunberg senza negare i cambiamenti climatici?"

25. JOE, "Slavoj Žižek Interview: Why I Like Greta Thunberg."

26. *Intercept*, "The Right to a Future, with Naomi Klein and Greta Thunberg."

27. Brockes, "When Alexandria Ocasio-Cortez met Greta Thunberg."

28. In Italian this high-speed train is known by the acronym TAV (Treno Alta Velocita).

29. On the NO TAV conflict, see CDCA, "NO-TAV Movement against High Speed Train, Val di Susa, Italy."

30. I wish to thank Professor Maria Federica Palestino who has worked for a year to involve high school students and teachers in our research activities. See Palestino, Quagliano, and Vetromile, "Pupils at the Forefront."

31. Fridays For Future Italia, "Report Assemblea Nazionale di Napoli."

32. Giovanna DiChiro, "Acting Globally," 234.

33. A counterproof of this argument comes from the more traditional Marxist organization Sinistra Classe Rivoluzione (Left, Class, Revolution), which expresses its critique of that FFF final document. For this organization, the national assembly, in searching for a compromise among the diverse areas of the movement, failed to embrace an explicitly anticapitalist agenda (Sinistra Classe Rivoluzione, "Non un passo Avanti").

34. Working Group on Practices, "Report."

35. The sociologist Nicholas Dines defined the *centri sociali* (social centers) as "urban spaces—generally disused factories or abandoned state property—occupied by groups of mostly young people who reuse them for political, social and/or cultural activities" ("Centri sociali," 90). Politically, the centri sociali are characterized by a radical anti-capitalist agenda and antagonist practices.

36. Battino, "Assedio alla gabbia del Casteller."

References

Armiero, Marco. "Seeing Like a Protester: Nature, Power, and Environmental Struggles." *Left History* 13, no. 1 (2008): 59–76.

Armiero, Marco. *A Rugged Nation: Mountains and the Making of Modern Italy*. Cambridge: White Horse, 2011.

Armiero, Marco. "Garbage under the Volcano. Fighting for Environmental Justice in Naples, Italy, and Beyond." In Armiero and Sedrez, *History of Environmentalism*, 167–84.

Armiero, Marco. "Is There an Indigenous Knowledge in the Urban North? Re/Inventing Local Knowledge and Communities in the Struggles over Garbage and Incinerators in Campania, Italy." *Estudos de Sociologia* 20, no. 1 (2014). https://www.periodicos.ufpe.br/revistas/revsocio/article/view/235511/28498.

Armiero, Marco, and Giacomo D'Alisa. "Rights of Resistance: The Garbage Struggles for Environmental Justice in Campania, Italy." *Capitalism Nature Socialism* 23, no. 4 (2012): 52–68.

Armiero, Marco, and Massimo De Angelis. "Anthropocene: Victims, Narrators, and Revolutionaries." *South Atlantic Quarterly* 116, no. 2 (2017): 345–62.

Armiero, Marco, and Lise Sedrez, eds. *A History of Environmentalism. Local Struggles, Global Histories*. London: Bloomsbury, 2014.

Armiero, Marco, and Leandro Sgueglia. "Wasted Spaces, Resisting People: The Politics of Waste in Naples, Italy." *Revista Tempo e Argumento* 11, no. 26 (2019): 135–56.

Battino, Anna Irma. "Assedio alla gabbia del Casteller: Smontate centinaia di metri di rete." *Global Project*, October 18, 2020. https://www.globalproject.info/it/in_movimento/assedio-alla-gabbia-del-casteller-smontate-centinaia-di-metri-di-rete/23051.

Bonora, Stefano. "Greta e i . . . gretini." *Sinistrainrete*, April 21, 2019. https://www.sinistrainrete.info/ecologia-e-ambiente/14814-quattro-interventi-sul-fenomeno-greta.html.

Brockes, Emma. "When Alexandria Ocasio-Cortez met Greta Thunberg: 'Hope is Contagious.'" *Guardian*, June 29, 2019. https://www.theguardian.com/environment/2019/jun/29/alexandria-ocasio-cortez-met-greta-thunberg-hope-contagious-climate.

Capone, Nicola. "The Assemblies of the City of Naples: A Long Battle to Defend the Landscape and Environment." *Capitalism Nature Socialism* 24, no. 4 (2013): 46–54.

Capone, Nicola, Antonella Cuccurullo, and Flora Micillo. *Allarme rifiuti tossici. Cronaca di un disastro annunciato* (*Toxic Waste Warning. Chronicles of a Foretold Disaster*). Naples: Palazzo Marigliano, 2006. https://www.allarmerifiutitossici.org/rifiutitossici/articles/art_4.html.

Cavallini, Annalisa. "Trans Adriatic Pipeline (TAP) in Puglia, Italy." *Environmental Justice Atlas*, August 18, 2019. https://www.ejatlas.org/conflict/gasdotto-trans-adriatico-tap-trans-adriatic-pipeline.

CDCA. "NO-TAV Movement against High Speed Train, Val di Susa, Italy." *Environmental Justice Atlas*, August 18, 2019. https://www.ejatlas.org/conflict/no-tav-movement-against-high-speed-train-val-di-susa-italy.

Chakrabarty, Dipesh. "Of Garbage, Modernity, and the Citizen's Gaze." *Economic and Political Weekly* 27, nos. 10–11 (1992): 541–47.

Croce, Benedetto. *Un paradiso abitato da diavoli*. Edited by Giuseppe Galasso. Milan: Adelphi, 2006.

D'Agostino, Peter. "Craniums, Criminals, and the 'Cursed Race': Italian Anthropology in American Racial Thought, 1861–1924." *Comparative Studies in Society and History* 44, no. 2 (2002): 319–43.

DiChiro, Giovanna. "Acting Globally: Cultivating a Thousand Community Solutions for Climate Justice." *Development* 54, no. 2 (2011): 232–36.

Dines, Nicholas. "Centri sociali: Occupazioni autogestite a Napoli negli anni novanta." *Quaderni di Sociologia* 21 (1999): 90–111.

Fagan, Pierluigi. "In direzione ostinata e contraria." *Sinistrainrete*, April 21, 2019. https://www.sinistrainrete.info/ecologia-e-ambiente/14814-quattro-interventi-sul-fenomeno-greta.html.

Fisher, Ian. "In Mire of Politics and the Mafia, Garbage Reigns." *New York Times*, May 31, 2007. https://www.nytimes.com/2007/05/31/world/europe/31naples.html.

Fridays for Future Italia. "Report Assemblea Nazionale di Napoli." https://www.fridaysforfutureitalia.it/report-2-assemblea-nazionale/.

Il simplicissimus. "Timeo Gretas et dona ferentes." https://www.sinistrainrete.info/ecologia-e-ambiente/14814-quattro-interventi-sul-fenomeno-greta.html.

Intercept. "The Right to a Future, with Naomi Klein and Greta Thunberg." September 10, 2019. https://www.theintercept.com/2019/09/06/greta-thunberg-naomi-klein-climate-change-livestream/.

JOE. "Slavoj Žižek Interview: Why I Like Greta Thunberg." YouTube, September 26, 2019. https://www.youtube.com/watch?v=IaF2_rDOLw8.

Iovino, Serenella. "Naples 2008, or, the Waste Land: Trash, Citizenship, and an Ethic of Narration." *Neohelicon* 36, no. 2 (2009): 335–46

Martínez Alier, Joan, and Ramachandra Guha. *Varieties of Environmentalism: Essays North and South*. London: Earthscan, 1997.

Nadeau, Barbie. "Naples Suffers as Garbage Piles Up." *Newsweek*, October 27, 2010. https://www.newsweek.com/naples-suffers-garbage-piles-73753.

Nixon, Rob. *Slow Violence and the Environmentalism of the Poor*. Cambridge, MA: Harvard University Press, 2011.

Palestino, Maria Federica, Simona Quagliano, and Elena Vetromile. "Pupils at the Forefront: The School-Work Interchange on Climate Change between University and High School in Naples." *Undisciplined Environments*, July 10, 2019. https://www.undisciplinedenvironments.org/2019/07/10/pupils-at-the-forefront-the-school-work-interchange-on-climate-change-between-university-and-high-school-in-naples/.

Pieterse, Edgar. *City Futures: Confronting the Crisis of Urban Development*. London: Zed Books, 2008.

Redazione. "Si può criticare Greta Thunberg senza negare i cambiamenti climatici?" *Sinistra*, November 16, 2019. https://www.sinistra.ch/?p=8253.

Schlosberg, David, and Lisette B Collins. "From Environmental to Climate Justice: Climate Change and the Discourse of Environmental Justice." *WIREs Climate Change* 5, no. 3 (2014): 359–74.

Sinistra Classe Rivoluzione. "Non un passo avanti—Un bilancio della seconda assemblea nazionale di Fridays for Future." October 21, 2019. https://www.rivoluzione.red/non-un-passo-avanti-un-bilancio-della-seconda-assemblea-nazionale-di-fridays-for-future/.

Smoltczyk Alexander, Andreas Ulrich, and Andreas Wassermann. "How the Mafia Helped Send Italy's Trash to Germany." *Spiegel International*, March 4, 2010. https://www.spiegel.de/international/europe/the-garbage-of-naples-how-the-mafia-helped-send-italy-s-trash-to-germany-a-681469.html.

Sohn-Rethel, Alfred. "Das Ideal des Kaputten: Über neapolitanische Technik." Reprinted in *Salad of Pearls* (blog), November 3, 2018. https://www.saladofpearls.wordpress.com/2018/11/03/the-ideal-of-the-broken-down-on-the-neapolitan-approach-to-things-technical-1926-by-alfred-sohn-rethel.

Stop Biocidio. "Piattaforma contro il biocidio, la devastazione ambientale, i roghi." *Global Project*, July 23, 2017. https://www.globalproject.info/it/in_movimento/stop-biocidio-piattaforma-contro-il-biocidio-la-devastazione-ambientale-i-roghi/20977.

Venn, Couze. "Rubbish, the Remnant, Etcetera." *Theory, Culture, and Society* 23, nos. 2–3 (2006): 44–46.

Verdicchio, Pasquale. "The Preclusion of Postcolonial Discourse in Southern Italy." In *Revisioning Italy. National Identity and Global Culture*, edited by Beverly Allen and Mary Russo, 191–212. Minneapolis: University of Minnesota Press, 1997.

Viggiano, Laura, and Catherine Hornby. "President Piles Pressure on Italy Government over Naples Trash." *Reuters*, June 24, 2011. https://www.reuters.com/article/us-italy-rubbish/president-piles-pressure-on-italy-government-over-naples-trash-idUSTRE75N20Q20110624.

Working Group on Practices. "Report." From the Fridays for Future national assembly, Naples October 5–6, 2019.

Zhok, Andrea. "Greta e il problema." Sinistrainrete, April 21, 2019. https://www.sinistrainrete.info/ecologia-e-ambiente/14814-quattro-interventi-sul-fenomeno-greta.html.

Rooting Out Injustices from the Top

The Multispecies Alliance in Morro da Babilônia, Rio de Janeiro

Lise Sedrez and Roberta Biasillo

Morro da Babilônia (Babilonia Hill), a favela near the famous Copacabana Beach in Rio de Janeiro, Brazil, is the site of a successful decade-long project of urban reforestation. The project results from an alliance between community, local government, and a large retail center in the area. After thirty years of active efforts, the results are impressive not only regarding the blossoming of a young forest and the return of native biodiversity but also regarding the pride and the reshaping of identity that took place within the community.

The need for reducing the environmental vulnerability of the area emerged in the 1970s and 1980s when a series of bushfires threatened the residents' homes and economic activities. In 1989 the Cooperativa de Reflorestadores da Babilônia (CoopBabilônia)[1] was established, and since 1994 local officers have implemented and financially and technically supported a reforestation project. Besides bushfires, what really boosted this severe socioecological transformation was the need to prevent soil erosion and the growing of the favela in landslide-prone area.[2] Soil consumption and informal urbanization of any available space reached their peak in 1980–90—when the overall city growth rate was 7.6 percent while favela populations surged by 40.5 percent—and decreased from 1990 to 2000 when the city's growth rate leveled off at less than 7 percent. Yet favela populations continued to grow by 24 percent.[3]

A historical interpretation through the lens of climate justice of the favela's transformative process is timely, if we consider that the floods between 1995 and 2004 affected 560,000 people in South America, and between 2005 and 2014 this number increased approximately four times,

DOI 10.1215/01642472-9495132

rising to 2.2 million.[4] More specifically, Brazil has become the only Latin American representative among the ten most affected countries in the world by weather- and climate-related disasters, with an absolute number of fifty-one million affected people between 1995 and 2015; Brazil has the second-highest loss potential by flooding in the world ranking of emerging countries.[5]

This has a very concrete meaning for the residents of Morro da Babilônia. While powerful storms and floods are part of the history of Rio de Janeiro, as in most coastal tropical cities, there is a direct connection between the increase of oceanic temperatures and the frequency of coastal rainstorms.[6] Rio de Janeiro is at least 1.5°C warmer than a hundred years ago, according to climatologist Carlos Nobre, and frequent extreme rainstorms are a new reality, confirmed yearly by record-breaking storms.[7] More than rainfall values, new extreme rainstorms bring about a fear of loss of lives and livelihoods for the Morro da Babilônia community.

This article explores past and present socionatures of Morro da Babilônia and establishes a connection between community and landscape transformations. In the first section, our interviews with workers, residents, and public officers who played an important role in the reforestation project, from both the community and the government, propose that the transformation did not bring back a previous landscape. On the contrary, it has rather produced a new patch of Atlantic rainforest in which so-called invasive species (Northern marmosets and Asian jackfruit trees) are part of a single biome together with people, makeshift houses, vultures, *pitangas* (Brazilian cherries), and boa constrictors. At the same time, the oral history project reveals the implications of presenting the current favela as a development of a former *quilombo*, a term that carries memories of slavery, suggests the presence of an original forest, and offers a window of opportunity to legalize the presence of the community in that area.

The second section focuses on how the parallel transformation of community and environment denaturalizes climate and climate change adaptation and mitigation policies. Adaptation to climate change merges with adaptation to political discourse that values climate change–related policies. While the concerns about soil erosion and loss of livelihood are recent and tangible, they evoke older fears of loss and eviction from the area. More importantly, the way in which community leaders chose to confront these new dangers conforms to previous patterns of resistance to older threats. Reforestation, together with its cultural meaning for the community and its relevance for mitigation policies, is the landscape transformation that fixes topsoil to rocks, avoiding landslides, and anchors the community of stable but informal residents to the place.

Finally, in the conclusion of the article, we discuss the relations of the favela leaders and the research team. Informers, as much as the team, demonstrate a key interest in the research. Far from being passive subjects, they sought to negotiate the design of the research project in order to use it as a tool to support their rights to the city, and their right to that particular place in the city. Creating a forest and developing a narrative are both survival strategies for the community. Just like the narrative of the origins of the favela allows its residents to claim a deep-rooted legitimacy for their residency in that place, the narrative of reforestation supports their claims of it being a *favela ecológica* (an ecological slum), as historian Camila Moraes named the Morro da Babilônia.[8]

Following other recent studies, we problematize the construction of sociopolitical and symbolic marginalization of the dwellers of the favelas in Rio de Janeiro[9] and, more specifically, we argue for the possibilities that the climate change context offers to redefine social and ecological protagonists. In this perspective, we highlight the relationship between the city and its residents—encompassing both upland and lowland districts—and the forest and its nonhuman inhabitants. How ecological elements and human actors merged together in Morro da Babilônia exemplifies the potential of a multispecies alliance in places that carry the scars of century-old legacies of socioeconomic inequalities. The favela's young forest proves that "these places can be lively despite announcements of their death" and in a state of social precarity and climatic vulnerability, its residents "don't have choices other than looking for life in this ruin."[10] The wooded area, far from being an original forest, can be considered what environmental humanities scholar Anna Tsing has defined as "third nature," namely an unexpected formation characterized by temporal polyphonies, human and nonhuman alliances, assemblages of institutional and grassroots actors, combinations of hope and despair.[11] Indeed, this article shows how the forested favela turned out to be an unexpected habitat for new specimens of native and nonnative plants and animals and for economic activities run by residents. The production of such a third nature represents yet another empowering factor for the community since it calls back indigenous framings of the world as a plural reality where every being is connected to others, in a complex and instigating biotic landscape.[12]

This is an orally informed research conducted by Natasha Barbosa, Letícia Batista, and Lise Sedrez that acknowledges history as "a means of addressing issues of social justice by empowering marginalized groups 'without history' both to create a heritage narrative by and for the people and to insert these multiple narratives into mainstream national origin stories and provide them with a role in these."[13] By adopting an oral- and environmental-history approach, this article places the recent climate injustice within a larger context of long-lasting social and spatial differences tracing its origin back to early twentieth-century power relations.

The Past Was Not a Walk in the Park

The Inter-American Development Bank provided $180 million in funding for the Favela-Bairro (Slum-Neighborhood) project in Rio de Janeiro in 1995. The project sought to integrate existing favelas into the fabric of the city through infrastructure upgrading and service increases. The Favela-Bairro project involves 253,000 residents in 73 communities.[14] Successful aspects of this large project were a committed and flexible city government and the use of intra- and extrainstitutional partnerships with NGOs, the private sector, churches, and the general population. Especially instrumental was the use of grassroots-level infrastructure upgrading experts as project managers who could work easily with both the government and community members.[15] Despite significant improvements in consumption of collective urban services, household goods, schooling, and social mobility, the stigma of place and race, the increase in unemployment, and the inability to translate educational gains into concomitant income or occupational gains remained[16] and coexisted with other forms of development and the risks of social exclusions due to the "touristification" and gentrification.[17] When a similar project, Projeto Mutirão Reflorestamento (Collaborative Reforestation Project), came to Babilônia, the community hoped for income alternatives in an area where dangerous criminal groups preyed on local youths. The planting and tending of the seedlings were, after all, paid for by the municipality and were a reliable and steady, if not overgenerous, income. The collaboration with the municipality was intended to transform a long-lasting socionature.[18]

The interaction between any favela and the state is always fraught with ambiguities. Land ownership is seldom clear; rents and sales of houses take place as informal transactions with little regulation; bank loans or mortgages are mostly unheard of. Often, favela residents are squatters, and they occupy public land in areas unmarked for urbanization for myriad reasons. Although public housing projects from the late twentieth century share many characteristics with favelas, the classic shantytown is made of makeshift houses, sometimes with multigenerational families, and sometimes with immigrants from poorer regions of the country recently arrived to a city with a large housing deficit and inconsistent housing policies. Urbanization services are irregular and often hard-won conquests after decades-long struggles with public authorities. Frequently, the presence of the state is limited to a repressive apparatus, which aims to contain crime within the favela or to contain the expansion of the favela itself. In fact, most favelas in Rio de Janeiro have unforgiving histories of resisting and experiencing forcible removal. Some were completely erased, such as the Praia do Pinto and the Catacumba, which were once located in the wealthy neighborhoods of Leblon and Ipanema, and were finally removed in the 1960s. Morro da Babilônia is no exception.

Favelas are thus vulnerable spaces—socially, politically, and environmentally. Community survival is as threatened by the ever-present menace of forcible removal as it is by landslides and storms. Ironically, the rationale for removal, as presented by the government institutions, particularly since the 1960s, was based on the danger of landslide and flood hazards to which these communities are exposed. This policy was somehow halted in the 1980s but not terminated. However, as environmental protection gained relevance after the 1992 Earth Summit, the rhetoric of favelas as cradles for petty crime and drug lords and as synonyms for urban expansion at the expense of green areas replaced previous arguments about their susceptibility to hazards. Nurturing and cultivating a forest helped challenge these narratives, so that each new tree planted contributes to securing the community's roots.

Many of those we interviewed were not themselves native to the Morro da Babilônia—nor to the city of Rio de Janeiro. Brazil underwent a powerful urbanization process in the mid-1950s, and by the 1970s more than 50 percent of the Brazilian population lived in cities. Migrant waves of poor, rural workers fed into the large Brazilian cities, including Rio. The hilly area, near the affluent and expanding Copacabana with its strong demand for unskilled labor (maids and construction workers, for instance) was a beacon for the new arrivals. Most of the residents of Morro da Babilônia, including the community leaders, are themselves immigrants or children of immigrants who arrived in the mid-1950, 1960s, or even 1970s. Among all the interviews with residents our group conducted (more than forty people), only two or three claimed to be residents of Morro da Babilônia for generations, and none recalled a forest standing in what is today the reforested area. As children, they used that particular area for running, playing soccer, and flying kites—all activities common for boys in Brazil, but in treeless areas. The women we interviewed, like D. Maria Teresa or D. Maria Elizabete, told us those were areas where they would not go alone as girls, for fear of the very tall grass, where (it was implicit) dangerous men could hide—but none told us of fear of a dark forest or tall trees.[19] It was just not in their memory. But what was their background knowledge of the area? We asked our informers what they knew about the past of the place where they have lived and what they had heard the elderly tell them. It was a *quilombo*, Palô, a community leader, told us.[20]

Quilombo is a powerful word in Brazilian history. It refers to a community of runaway enslaved people, at least until 1888, when Brazil abolished slavery, becoming the last country in the Americas to do so. Even after that, *quilombo* may refer to communities of freedmen who refused to serve their former masters and sought to build relatively autonomous communities. More importantly, the Brazilian constitution of 1988 recognized the right to land ownership by the *quilombolas*, the descendants of slaves who live in the communities their ancestors formed, even when there is little legal documentation to back their claim.

Is it true? Is Morro da Babilônia a former quilombo? Palô speaks with firm conviction, but there is no corroborating evidence. He was born in Morro da Babilônia, but his mother, Dona Persília, arrived in Rio de Janeiro as a young bride, probably by the mid-1940s, from Minas Gerais. Dona Persília was herself an important community leader and was several times the president of the neighborhood association. She was also the first person to seek allies in the wealthier neighborhood to document the vulnerable sites of Morro da Babilônia, the rocks that could fall on the top of the houses during a strong rainstorm or a landslide.[21] She knew everybody who lived there over five decades, and we can suppose Palô talked to many of the elderly dwellers of the Morro da Babilônia. Maybe he heard this story from one of them, and perhaps that person was a descendant of former slaves.

Palô's claim that Morro da Babilônia was a former quilombo, however, introduces further implications. First, the term defines the Morro da Babilônia as a Black community, not only for the racial makeup of its current dwellers, but historically so. By evoking a past of Black resistance, it unites part of the community and celebrates its roots. A revigorated Black movement has become more active in the neighborhood association since 2000 and has promoted local heritage through several activities, such as the teaching of African drums to the youth. In the forested area, there is a small grove known as *área da macumba* where Afro-Brazilian religious rites are practiced. This revival has been endangered in recent years by an alliance, in many favelas, between Pentecostal Christian churches and drug dealers, which directly threatened the Afro-Brazilian priests and priestesses, sometimes forcing their eviction.[22] Nevertheless, the quilombo past remains a unifying narrative, even if the religious aspects of the African heritage are shunned by some conservative Christians.

Second, the quilombo draws a claim of continuity of occupation of the area, with legal standing to request proper ownership of the land. This is not a negligible factor, as the history of favelas in Rio in the twentieth century is also a history of attempts by the government to remove poor residents from these areas whenever the interests of the wealthier part of the city demand so. In the absence of proper land titles, community ownership is acceptable under Brazilian law only in certain circumstances, as in the case of former quilombos. There have been no attempts to formally register Morro da Babilônia as a quilombo legacy, but bringing up this aspect in itself leverages considerable political power. At the end, it is one of the above-mentioned strategies of survival, as it can strengthen the aura of legitimacy over the place and stave off potential threats of forcible removal.

Finally, because quilombos are often described as secluded communities in forested areas,[23] this expression evokes the memory of an original forest, even if neither Palô nor any of his childhood friends and neighbors can actually recall that forest. These features align with the description of the elders of how to build houses in the region. One first finds a place, clears the area, and then uses the trunk of the felled trees to build the structure of the house. The foundations are not very deep—the shallow topsoil covers a granite formation. Mud and grass are then used to fill the gaps. This technique is used all over Brazil, probably even before the arrival of the Europeans. There are few houses like this in Morro da Babilônia today, very likely dated back to the 1950s. The modern dwellings, even the most precarious ones, use concrete and brick—urban materials, bought in stores and produced in factories.

The context can be reconstructed also via two different corroborating sources. The first one is more famous: Marcel Camus's movie *Black Orpheus* (1959) was filmed in Morro da Babilônia and offers a fair overview of its houses, streets, and general surroundings. The gorgeous view that is still a source of pride and pleasure for the residents has also provided them with visual records of their past, when so many other poor communities in Rio de Janeiro cannot boast the same good fortune. Beauty, therefore, is part of the collective memory and the identity.

The second corroborating source emerged from archival collection where we looked for pictures and descriptions of the Morro. One of the first references we found was in the newspaper *Correio da Manhã*. A 1907 news article, illustrated with photographs of the Morro da Babilônia on the front page, describes in detail how the poor were occupying the hills of Rio de Janeiro, and Morro da Babilônia was the chosen example. The makeshift houses, with mud, clay, grass, and the occasional tin roof, were present. The residents had small vegetable patches of land, where they had cleared the forests. They were *caboclos* (a mix of native Brazilians and Europeans) and Blacks; they were former soldiers, living rent-free in the hills, and they would go to "to the city down below" to sell small objects they produced or to hire out for the occasional job.[24]

In these documents of the early twentieth century, the forest, the same forest Palô and N.,[25] another worker in the project, cannot recall, is present. But not in the way the recent reforestation project shaped it. It is there as a "strong," "overwhelming," and "untouched" presence together with informal settlements at least since the early nineteenth century. The forest offered, then, a green barrier that protected the poor residents, in an almost "out of the grid" experience, as described in the *Correio da Manhã*.[26]

Other articles from the same newspaper were less generous. They depicted how criminals hid in the "glorious greenness" of the area, bringing danger and chaos to a place that was remarkable for its beauty.[27] The beauty of nature was in contrast with the ugliness of the poor residents, who cut trees and built ugly houses. The hill was named probably after the Suspended Gardens of Babilônia, for its beauty, a natural wonder in Rio de Janeiro, and it was threatened by uncivilized thugs. In 1913, the newspaper *O Paiz* celebrated that a "pack of female thieves" who lived in the Morro da Babilônia, under the command of the Black woman Olga, had finally been arrested.[28] In 1915, residents of the distinct neighborhood of Leme begged the police to do something about the "pack of good-for-nothing men" who used to meet in the Morro and threaten the "respectable citizens" of Leme.[29] In the past, as in the present, the conflicting portraits of the favela marked its troubled relationship with the state and with its wealthier neighbors—and the ever-present requests for its forcible removal.

By bringing up these memories of ecologies, we are forced to question the concept of reforestation. The original forest has a history of interaction with human societies of the area but a history that has changed over time and that can be evoked but cannot be simply restored. It has continuously been re-created under new parameters of coexistence among the different groups, humans and nonhumans. Certain species, such as tall grass and jararacas (*bothrops jararaca*, a poisonous viper), increased their presence in the region during most of the twentieth century, when the tree area receded. Later on, in the last thirty years, with the growth of the forestation project, they were replaced by bush and tree vegetation and returning boa constrictors, which are a good indicator of a healthy Atlantic forest. This indigenous species was also joined by nonnative fauna and flora, such as marmosets (originally from the state of Bahia) and jackfruit trees (originally from Asia). The new biome cannot be uprooted.

In its several versions, these ecologies of memories bring up a forest of quilombos, a forest of beauty similar to the wonders of the old world, a forest of thieves and criminals, a forest of protection for poor former soldiers and minorities, a forest that appeared in postcards, a forest that provided materials for houses and small animals for urban hunters, a forest replaced by cleared areas where children play, and where trees were forgotten, a forest that offered hope. The re-created memory of race and nature, of the quilombo and a "pristine" forest, nurtures the modern-day experience of resistance of the mostly Black residents of Babilônia and their role in planting new forests.

It is within this context of memory that we must understand the multispecies coexistence space in Morro da Babilônia in the early twenty-first century.

Beyond Loss and Damage

Recently, climate change is taking us back to the very beginning of the story of this community and place; it is showing us the mud and shallow topsoil covering the hills around the downtown area, and it is putting slums and rich neighborhoods once again in connection, both though disasters and mitigation projects. And yet, climate change is pushing institutions to welcome and support sustainable actions and initiatives.

Since 1986, the Environmental Secretariat of Rio's City Hall (SMAC) has led a community reforestation program, the above mentioned Projeto Mutirão Reflorestamento, and planted more than six million seedlings on 2,200 hectares of land within the city limits. Rio had long suffered from deforestation of its hills as a result of development, causing soil erosion, sediment buildup in waterways, floods, landslides, and pools of water filled with disease-carrying mosquitos.[30] After the massive flood that hit Rio in 2010, the municipality, together with the World Bank, started to develop a strategic plan to tackle climate change effects: the Rio de Janeiro Low Carbon City Development Program (LCCDP) sought to help the city government identify and finance climate change mitigation opportunities across a number of urban sectors. Among these actions, Rio's community reforestation program proved so successful that a new urban forestry project called Rio Capital Verde (Rio Green Capital) was launched. Rio Green Capital plants trees in remote areas of Rio, and the project aims to earn carbon credits under the Rio Low Carbon City Development Program (LCCDP).[31]

In addition, climate change is making us historians look at and forward different stories. Interviews with urban disaster victims help us further explore the newly formed forest, which has created a different socionature in the midst of dramatic loss and damage. The so-called traveling soils of Rio de Janeiro hills created risky areas for living, especially as deforestation took place in the city. Floods and landslides have made clear how poor population's social and political vulnerability would be followed by environmental vulnerability as well. It was not only a matter of risks concerning diseases, lack of good sanitary conditions, bad sewage systems, and lack of water supply. Owing to floods and landslides, their immediate present would become uncertain. Survival in their communities, and even physical survival, would not be guaranteed. Those communities were aware of the risk of torrential rains. When floods were reasonably bearable, they would become part of poor people's everyday life. In these cases, torrential rain would be seen as just another occasional providential challenge to be accepted.[32] However, torrential rain could often be strong enough to blur the limits of what was bearable and what was not. When that happened, previous social experiences would lose their meaning, and what used to be an invisible everyday precariousness would become an evident crisis.[33]

If for some institutional actors the forest represents primarily a means for strengthening urban resilience through nature and an ecosystem-based measure reducing landslide risk in Rio de Janeiro,[34] for the residents it represents a complex ecosocial infrastructure. While the forest ultimately protects them from climate extreme events, it also offers visibility and socioeconomic means of survival—and, strategically, political protection against the depiction of the favelas as villains in the Rio Green Capital narrative.[35]

In many ways, the Morro da Babilônia residents trust their nonhuman partners more than they trust their institutional partners. In our interview with Claudia França and Luiz Lourenço, two public servants at the SMAC who were among the pioneers in the reforestation project, their own skepticism regarding the government's investments in the project was noticeable. Governments, according to França and Lourenço, have a short attention span. The survival of the reforestation project is nothing short of a miracle—and the result of painstaking efforts by SMAC's technical personnel and the residents of the communities. In fact, in other areas with little visibility the project was interrupted, regardless of its positive environmental impact. The Projeto Mutirão Reflorestamento managed to survive different administrations for decades just because it was relatively inexpensive and because it has received many international awards and acknowledgments.[36] Mayors saw in the project a way to increase their leverage in international carbon credit negotiations, and a charming billboard to Rio de Janeiro's "green vocation." In addition, no mayor wanted the political costs of shutting down the project. Thus, França and Lourenço encouraged the communities to claim the reforestation project as their own because too much of its success had depended on the executive branch, and mayors change more rapidly than the seedlings grow.

Forest workers such as Palô and N. are perfectly aware of the fickleness of *carioca* politicians. Since the project began in the 1990s, the favela leaders have sought different alliances and partnerships: with the powerful commercial shopping center Rio Sul, with the wealthier neighbor associations of Leme and Copacabana, with ecotourism agencies, and, eventually, with the university—as in embracing our research. They seek to strengthen their networks to make sure that the political costs to shut down the project would be steep for any mayor. As we argue, they saw these alliances as necessary steps to control their own narratives—and as part of a key strategy of community survival.

While we highlight the importance of narrative control, however, we must also examine the less planned consequences of this strategy. In the process of securing alliances, income and state presence in Morro da Babilônia, the favela dwellers reaffirmed their connection to the place where they live and their nonhuman partners. First, they have invested in knowledge. Palô recalls how little he knew about planting in those early years and how his body hurt after one day planting seedlings under the sun. N. was drafted by Palô to the project when he was fifteen years old, and already with some experience in the few and dangerous job opportunities left for young Black men in favelas, often connected to drug trafficking. Many of his former childhood friends have died violently, while he sees a different path in front of him. He has probably less than four years of formal schooling and is currently responsible for teaching visitors about the impact of the forest on the local microclimate. His eyes shine when he goes on about forest biodiversity and how the experience has changed his life.[37]

Secondly, they see the nature around them, their environment, as a point of pride and as part of their new identity. D. Maria Elizabeth claims she breathes a different air, "fresher, cleaner, purer," when she returns home from her work downtown.[38] Many of our informers invited us to take time to enjoy the magnificent view before starting the interviews. Others pointed out the beauty of the Brazilian cherry trees, the colorful birds, and the exuberance of fruit and green in the heart of the megacity.

Thirdly, the participants in the reforestation project present the forest as a business card to enhance their relationship with the city at large. It is common for any of the workers to pull out their phones when they go out to work with the trees to capture a particularly beautiful flower or to seek out marmosets, anteaters, boas, or other animals common to the Atlantic Rainforest with their cameras. They continuously update portfolios about the project and document the transformation of the area, which are important tools to impress potential partners and sponsors. By recording and protecting their nonhuman partners, Palô and the other community leaders were able to develop connections with local schools, which frequently visit the area for environmental education activities; birdwatcher societies; urban trail hikers; and tourist agencies, which send customers and visitors to the well-marked ecotourist trails.

Environmental tourism and environmental education activities, centered on the new forest area, have helped to achieve a major victory for Morro da Babilônia, in the process redefining its place in the city. By providing visible ecological services, the favela residents reinforce their claim to legitimacy and full citizenship—their right to the city.[39] They thus undermine charges that the favelas destroyed the beauty of the forested hills, depreciated valuable real estate nearby, and presented inherent risk to the *favelados* themselves—and therefore must be removed by the state.

The interviews we conducted are also part of this larger process in which the community is eager to forge alliances but also to retain some control over its own image. When we first contacted the Cooperativa, in early 2018, our research team received a warm welcome but also some mistrust. We were not the first academic researchers to reach out to the community. Some had neglected to give back to the favela the results of their research and left behind a perception of academic extractivism. Our first months in Morro da Babilônia were basically a negotiation of trust, and we decided to remake the original design of the research to include some of the community's concerns. Eight months later, with interviews already concluded, Palô invited us to visit the planting area and meet with the workers on site. It was an entrancing experience, including about two hours climbing up the hill, among colorful flowers, seedlings, lizards. We saw two young boa constrictors, about four feet long, duly documented for PowerPoint presentations—it was a bit unsettling to see them so close to Copacabana beach. We reached a hilltop with sparse tree cover, and many black vultures—the area is in fact named after them, Morro do Urubu (Vulture Hill). For the delight of the research team, a sun-drenched pitanga tree (*Eugenia uniflora*, Brazilian cherry) was heavy with fruit. A makeshift soccer area, where some children from Morro da Babilônia played, was just beside the most amazing view of Guanabara Bay and the Sugar Loaf. The field trip marked a mutual commitment to work together from both the research team and the participants in the forestation project. In the next meeting, the research leader was encouraged to join the citizen council that advises in the management of the environmental protected area where the new forest is located, and we discussed plans for a future community ecomuseum.

Conclusions

This article demonstrates that environment and climate change can provide marginal groups with unexpected political leverage and can act as spaces of encounters and possibilities beyond the alternatives of degradation-restoration and beyond the ecological realm. Babilônia forest is the space where the downtown and uptown districts meet; where the municipality, grassroots initiatives, and international developmentalist actors collaborate; where legacies of the past, climate injustices, and social problems change trajectories and blend into an open-ended tale.

This unexpected assemblage, this third nature growing from spatial and social injustices is preaching to the choir of the "Multispecies City" in which collaborative practices of "learning with the non-human" provoke "changes in public discourse of who can be seen as part of the city."[40] In the 1990s, Brazil saw a transformation that Henry Acselrad defines as the "environmentalization of social demands and struggles": favela shanty towns entered the public and governmental agenda as "public problems" and the institutionalization involved also the residents, who started to create associations whose main goal was the reforestation of the slopes.[41] Whether this is actually enforced or remains in the discursive arena is still debatable, but certainly it depends significantly on how much the favela communities are aware of the high stakes and how much they are able to pursue their own agendas.

Communities incorporate climate change mitigation initiatives in their toolkit of strategies for long-term survival—survival as communities, as much as physical survival. However, in the case of Morro da Babilônia, while we conclude that the opportunities in the forestation project (such as income for local residents or bargaining power with the local authorities) fit within the traditional goals in this repertoire, the interviews also show how the development of the project itself has transformed the land, public servants, and community leaders as well.

Beautiful vistas notwithstanding, life in a favela is not a fairy tale, and climate change comes on top of many risks for vulnerable communities. In April 2019, Rio de Janeiro was hit by yet another extreme climate event: an extraordinary rainstorm—the strongest in the last twenty-two years—caused death and destruction all across the city. Ten people were killed, among them Doralice and Gerlaine Nascimento, buried in their home by a landslide that sank a portion of the favela where they lived, Morro de Babilônia. Climate change challenges do not diminish the magnitude of the other everyday struggles for favela dwellers. Our own research has two terrible bookends. In early 2018, a conflict between two heavily armed drug gangs prevented the research team from accessing the community for more than three months. In 2020, it was the COVID-19 pandemic that interrupted our work. Our research has also highlighted the unbalance of the multiple alliances developed by the community leaders. While most of our academic work has continued via home offices, this is not a viable alternative for those who must plant and tend to seedlings. Even if in better circumstances, favela residents still face rampant structural racism and threats of community removal by the city government. They are easy prey for organized crime and are often treated as second-class citizens by the police who should protect them from crime. Poverty and lack of economic opportunities combine with the frailty of education and health services, making life for men and women in favelas an everyday struggle. However, the experience of forestation in Morro da Babilônia suggests that confronting climate change may offer possibilities to tackle some of these other questions as well.

Lise Sedrez is professor of history at the Universidade Federal do Rio de Janeiro and a CNPq (Brazilian National Council for Scientific and Technological Development) researcher.She is one of the editors of *The Great Convergence: An Environmental History of BRICS* (2018).

Roberta Biasillo is assistant professor of contemporary political history at Utrecht University and visiting Max Weber Fellow at the European University Institute in Florence. She has coauthored *Mussolini's Nature: An Environmental History of Fascism* (forthcoming).

Notes

Support for work on this article was provided by FORMAS (Swedish Research Council for Sustainable Development) under the National Research Programme on Climate (contract 2017-01962_3) and by CNPq (Brazilian National Council for Scientific and Technological Development or Conselho Nacional de Pesquisa e Desenvolvimento Tecnológico).

1. CoopBabilonia's webpage is at coopbabilonia.blogspot.com (accessed March 1, 2021).
2. Lara, "One Katrina Every Year"; Brum and Lemos, "Urban Projects for Resilience."
3. O'Hare and Barke, "The Favelas of Rio de Janeiro;" Roy and Alsayyad, "Urban Informality," 107–108.
4. UNISDR, "The Human Cost of Weather-Related Disasters."
5. Gullo Cavalcante, Luz Barcellos, and Cataldi, "Flash Flood in the Mountainous Region of Rio de Janeiro State (Brazil) in 2011."
6. Aumann, Behrangi, and Wang, "Increased Frequency of Extreme Tropical Deep Convection."
7. Academia Brasileira de Ciências, "Tribunais fazem parte de uma nova realidade."
8. Moraes, "A invenção da favela ecológica," 461–63.
9. Perlman, *Favela*; Lannes-Fernandes, "The Construction of Socio-Political and Symbolical Marginalisation in Brazil"; Butler, "Transforming the Imaginary of Marginality."
10. Tsing, *The Mushroom at the End of the World*, 5–6.
11. Tsing, *The Mushroom at the End of the World*, vii–viii.
12. Viveiros de Castro, *Cannibal Metaphysics*; Kopenawa and Albert, *The Falling Sky*.
13. Weisman, "Oral Sources and Oral History," 57.
14. Gomes and Motta, "Empresariamento urbano e direito à cidade."
15. Inter-American Development Bank, "Brazil: Rio de Janeiro Urban Upgrading Program"; Inter-American Development Bank. *Programa de Urbanizacao de Assentamentos Populares do Rio de Janeiro*; Segre, "Formal-Informal Connections in the Favelas of Rio de Janeiro," 194–95.
16. Perlman, "The Metamorphosis of Marginality."
17. Freire-Medeiros, "A favela que se vê e que se vende"; Griffin, "Olympic Exclusion Zone"; Comelli, Anguelovski, and Chu, "Socio-Spatial Legibility, Discipline, and Gentrification." The Rio de Janeiro administration developed the 2016 strategic plan to ensure the good use of foreign investments coming to the city and to improve its governance. See Carloni, "Rio de Janeiro Low Carbon City Development Program."

18. Sedrez, Barbosa, "Narrativas na Babilônia."

19. Six women, collective interview, January 23, 2019. Names were changed as requested.

20. Carlos Antonio Pereira, Palô (community leader), interview, October 19, 2019.

21. Abílio Valério Tozini (president of the Leme Neighbourhood Association), interview, January 14, 2019. Dona Persilia's complete name was Persilia Pereira. As with Palô, we have opted to keep the names as they are known in the community.

22. Schipani and Leahy, "'Drug Traffickers of Jesus.'"

23. Quilombos were established in many different ecosystems in Brazil: forests, marshes, sandy regions, semiarid landscapes, etc. But the most famous quilombos, such as Palmares, in the eighteenth century, were in fact in forests, and that is how they remain in the Brazilian imagination.

24. *Correio da Manhã*, "No Morro da Babylonia," 1.

25. N., interview, December 16, 2019. Worker's name removed as requested.

26. *Correio da Manhã*, "No Morro da Babylonia," 1.

27. *Correio da Manhã*, "E continúa a derrubada das nossas florestas."

28. *O Paíz*, "A quadrilha da Babylonia," 6.

29. *Correio da Manhã*, "E continúa a derrubada das nossas florestas."

30. Braun, "Rio de Janeiro's Reforestation."

31. City of Rio de Janeiro, "GRI Report"; City of Rio de Janeiro, *Plano Estratégico da Prefeitura do Rio de Janeiro 2013–2016*; City of Rio de Janeiro and World Bank, "The Rio de Janeiro Low Carbon City Development Program"; Bittencourt et al., "Evaluating Preparedness and Resilience Initiatives."

32. This phenomenon has been analysed in Bankoff, *Cultures of Disaster*.

33. Maia and Sedrez, "Narrativas de um Dilúvio Carioca."

34. Lange, Sandholz, and Nehren, "Strengthening Urban Resilience through Nature."

35. Lima Gottgtroy de Carvalho, "O turismo no Morro da Babilônia."

36. Claudia França and Luiz Lourenço, interview, February 8, 2019.

37. N., interview, December 16, 2019.

38. Six women, collective interview, January 23, 2019. Names were changed as requested.

39. For the concept of the Right to the City, as employed by the authors, see Harvey, *Rebel Cities*.

40. Ernstson and Sörlin, "Toward Comparative Urban Environmentalism," 14–16; on the same topic, see also Myers, *Rethinking Urbanism*.

41. Acselrad, "Ambientalização das lutas sociais," 104.

References

Academia Brasileira de Ciências. "Tribunais fazem parte de uma nova realidade, diz climatologista." April 11, 2019. https://www.abc.org.br/2019/04/11/temporais-fazem-parte-de-uma-nova-realidade-diz-climatologista/.

Acselrad, Henry. "Ambientalização das lutas sociais: o caso do movimento por justiça ambiental." *Estudos avançados* 24, no. 68 (2010): 103–19.

Aumann, H. Hartmut, Ali Behrangi, and Yuan Wang. "Increased Frequency of Extreme Tropical Deep Convection: AIRS Observations and Climate Model Predictions." *Geophysical Research Letters* 45, no. 24 (2018): 13530–37. https://doi.org/10.1029/2018GL079423.

Bankoff, Greg. *Cultures of Disaster: Society and Natural Hazards in the Philippines*. Richmond, UK: Curzon, 2015.

Bittencourt, Bernardo K., Marcos P. Cannabrava, Trystyn K. Del Rosario, Michelle C. Hamilton, Molly E. Kampmann, Joseph T. McGrath, Bernardo B. Ribeiro, José Orlando Gomes, and James H. Lambert. "Evaluating Preparedness and Resilience Initiatives for Distressed Populations Vulnerable to Disasters in Rio de Janeiro, Brazil." *2013 IEEE Systems and Information Engineering Design Symposium* (2013), 58–63. https://doi.org/10.1109/SIEDS.2013.6549494.

Braun, Franka. "Rio de Janeiro's Reforestation Changes Life in the Favelas." *World Bank* (blog), September 9, 2013. https://blogs.worldbank.org/latinamerica/rio-de-janeiros-reforestation-changes-life-favelas.

Brum, Erika, and Maria Fernanda Lemos. "Urban Projects for Resilience in Flooding Areas." In *Rio de Janeiro: Urban Expansion and the Environment*, edited by José L. S. Gámez, Zhongjie Lin, and Jeffrey S. Nesbit, 113–23. New York: Routledge, 2019.

Butler, Udi Mandel. "Transforming the Imaginary of Marginality: Some Experiences from Rio de Janeiro." *Periferias* 1, no. 1 (2018). https://revistaperiferias.org/en/materia/transforming-the-imaginary-of-marginality-some-experiences-from-rio-de-janeiro/?pdf=311.

Carloni, Flavia. "Rio de Janeiro Low Carbon City Development Program." *GGBP Case Study Series* (2014). https://www.greengrowthknowledge.org/sites/default/files/downloads/best-practices/GGBP%20Case%20Study%20Series_Brazil_Low%20Carbon%20City%20Development%20Program%20Rio%20de%20Janeiro.pdf (accessed March 1, 2021).

Carvalho, Thays Lima Gottgtroy de. "O turismo no Morro da Babilônia (RJ): Do reflorestamento ao ecoturismo." *Revista Brasileira de Ecoturismo* 9, no. 1 (2016): 11–28.

Centre for Research on Epidemiology of Disaster and United Nations International Strategy for Disaster Reduction. *The Human Cost of Weather-Related Disasters 1995–2015*. Brussels–Geneva, 2015. https://unisdr.org/2015/docs/climatechange/COP21_WeatherDisastersReport_2015_FINAL.pdf.

City of Rio de Janeiro. *GRI Report Sustainability of Rio de Janeiro City Hall*. Rio de Janeiro: Rio de Janeiro's City Hall and Keyassociados, 2011.

City of Rio de Janeiro. *Plano Estratégico da Prefeitura do Rio de Janeiro 2013–2016—Rio mais integrado e competitivo*. Rio de Janeiro: Rio Prefeitura, 2012.

City of Rio de Janeiro and World Bank. *The Rio de Janeiro Low Carbon City Development Program—Program Document*. Washington, DC: World Bank, 2012.

Comelli, Thaisa, Isabelle Anguelovski, and Eric Chu. "Socio-Spatial Legibility, Discipline, and Gentrification through Favela Upgrading in Rio de Janeiro." *City* 22, nos. 5–6 (2018): 633–56.

Correio da Manhã. "No Morro da Babilônia." June 2, 1907, 1.

Correio da Manhã. "E continúa a derrubada das nossas florestas—O que se está praticando no morro da Babylonia é inacreditavel." December 29, 1915, 3.

Ernstson, Henrik, and Sverker Sörlin, "Toward Comparative Urban Environmentalism: Situating Urban Natures in an Emerging 'World of Cities.'" In *Grounding Urban Natures: Histories and Futures of Urban Ecologies*, edited by Henrik Ernstson and Sverker Sörlin, 3–53. Cambridge, MA: MIT Press, 2019.

Freire-Medeiros, Bianca. "A favela que se vê e que se vende: reflexões e polêmicas em torno de um destino turístico." *Revista Brasileira de Ciências Sociais* 22, no. 65 (2007): 61–72.

Gomes, Maria de Fátima Cabral Marques, and Motta Thaiany da Silva. "Empresariamento urbano e direito à cidade: considerações sobre os programas favela-bairro e morar carioca no Morro da Providência." *Revista Libertas* 13, no. 2 (2013): 54–79.

Griffin, Jo. "Olimpic Exclusion Zone: The Gentrification of a Rio Favela." *Guardian*, June 15, 2016. https://www.theguardian.com/cities/2016/jun/15/rio-olympics-exclusion-zone-gentrification-favela-babilonia.

Gullo Cavalcante, Marianna Rodrigues, Priscilla da Cunha Luz Barcellos, and Marcio Cataldi. "Flash Flood in the Mountainous Region of Rio de Janeiro State (Brazil) in 2011: Part I—Calibration Watershed through Hydrological SMAP Model." *Natural Hazards* 102, no. 3 (2020): 1117–34. https://doi.org/10.1007/s11069-020-03948-3.

Harvey, David. *Rebel Cities: From the Right to the City to the Urban Revolution*. London: Verso, 2019.

Inter-American Development Bank. Report of the Favela/Bairro Project "Brazil: Rio de Janeiro Urban Upgrading Program." Operation no. 898/OC-BR. Approved on November 1, 1995. (BR-0182).

Inter-American Development Bank. *Programa de Urbanizacao de Assentamentos Populares do Rio de Janeiro*. Washington, DC: Inter-American Development Bank, 1998.

Kopenawa, Davi, and Bruce Albert. *The Falling Sky. Words of a Yanomami Shaman*. Cambridge, MA: Harvard University Press, 2013.

Lange, Wolfram, Simone Sandholz, and Udo Nehren. "Strengthening Urban Resilience through Nature: The Potential of Ecosystem-Based Measures for Reduction of Landslide Risk in Rio de Janeiro." Lincoln Institute of Land Policy, June 2018. https://www.lincolninst.edu/sites/default/files/pubfiles/lange_wp18wl1.pdf.

Lannes-Fernandes, Fernando. "The Construction of Socio-Political and Symbolical Marginalisation in Brazil: Reflecting the Relation between Socio-Spatial Stigma and Responses to Violence in Rio de Janeiro." *International Journal of Humanities and Social Science* 4, no. 2 (2014): 53–67.

Lara, Fernando. "One Katrina Every Year: The Challenge of Urban Flooding in Tropical Cities." In *Rio de Janeiro: Urban Expansion and the Environment*, edited by José L. S. Gámez, Zhongjie Lin, and Jeffrey S. Nesbit, 18–26. New York: Routledge, 2019.

Maia, Andréa Casa Nova, and Lise Sedrez. "Narrativas de um Dilúvio Carioca: memória e natureza na Grande Enchente de 1966." *História Oral* 14, no. 2 (2012): 221–54. https://doi.org/10.51880/ho.v14i2.239.

Moraes, Camila. "A invenção da favela ecológica: Un olhar sobre turismo e meio ambiente no Morro Babilônia." *Revista Estud Sociologia* 18, no. 35 (2013): 459–74.

Myers, Garth. *Rethinking Urbanism. Lessons from Postcolonialism and the Global South*. Bristol, UK: Bristol University Press, 2020.

O'Hare, Greg, and Barke Michael. "The Favelas of Rio de Janeiro: A Temporal and Spatial Analysis." *GeoJournal* 56, no. 3 (2002): 115–240.

O Paíz. "A quadrilha da Babylonia." April 4, 1913, 6.

Perlman, Janice. "The Metamorphosis of Marginality: Four Generations in the Favelas of Rio de Janeiro." *Annals of the American Academy of Political and Social Science* 606, no. 1 (2006): 154–77.

Perlman, Janice. *Favela: Four Decades of Living on the Edge in Rio de Janiero*. Oxford: Oxford University Press, 2010.

Roy, Ananya, and Alsayyad, Nezar, eds. *Urban Informality: Transnational Perspectives from the Middle East, Latin America, and South Asia*. Lanham, MD: Lexington Books, 2004.

Schipani, Andres, and Joe Leahy. "'Drug Traffickers of Jesus' Drive Brazil Slum Violence." *Financial Times*, October 27, 2017. https://www.ft.com/content/b5096a18-b548-11e7-aa26-bb002965bce8.

Sedrez, Lise, and Natasha Barbosa. "Narrativas na Babilônia: Uma experiência de história oral, risco climático, reflorestamento e comunidade (1985–2015)." In *História oral e direito à cidade: Paisagens urbanas, narrativas e memória social*, edited by Andréa Casa Nova Maia, 79–98. Rio de Janeiro: Letra e Voz, 2019.

Segre, Roberto. "Formal-Informal Connections in the Favelas of Rio de Janeiro: The Favela-Bairro Programme." In *Rethinking the Informal City*, edited by Felipe Hernández, Peter Kellett, and Lea K. Allen, 163–180. New York: Berghahn Books, 2010.

Tsing, Anna L. *The Mushroom at the End of the World: On the Possibility of Life in Capitalist Ruins*. Princeton, NJ: Princeton University Press, 2015.

Viveiros de Castro, Eduardo. *Cannibal Metaphysics*. Translated and edited by Peter Skafish. Minneapolis: University of Minnesota Press, 2014.

Weisman, Brent R. "Oral Sources and Oral History." In *Encyclopedia of Global Archaeology*, edited by Claire Smith. New York: Springer, 2014. https://doi.org/10.1007/978-1-4419-0465-2_1401.

From the Occupied Parks to the Gardens of the Nation

Politics and Aesthetics of Urban Greenery in Post-Gezi Istanbul

Sinan Erensü, Barış İne, Yaşar Adnan Adanalı

On February 5–6, 2020, Istanbul Congress Hall hosted a first of its kind workshop on the future of the city's urban forests, registered trees, public parks, and recreational grounds. Aptly titled Istanbul Green Spaces Workshop, the event was organized by the Istanbul Metropolitan Municipality (IMM) and attracted some twelve hundred participants across fourteen parallel sessions on a variety of topics ranging from rooftop gardening to how to protect Istanbul's ancient *bostans* (market gardens). The main venue was packed in the morning of the first day. Municipal officers, landscaping professionals, academics, college students, and urban activists were all scrambling for seats to watch the opening address of the new mayor of Istanbul, Ekrem Imamoğlu. Only six months earlier, his election campaign had ended the Justice and Development Party (AKP) cadres' twenty-five-year-long control over the city government, a long political reign begun in 1994 by a then-rookie politician, now president, Recep Tayyip Erdoğan. Mayor Imamoğlu began his talk by hinting at the green record of the outgoing local leadership: "If we tell 50 random Istanbulites about this green spaces workshop," the mayor continued sarcastically, "most would react, 'What green space? Have they left any of it?' Sad, but true!"

The same day, the new green spaces vision of the new municipal administration was made public as well. Accordingly, the immediate goals would be increasing participation in planning and design, bringing more

DOI 10.1215/01642472-9495146

green space to the most disadvantaged neighborhoods, replacing wet landscaping with less water-intensive solutions and saving $25 million spent annually on the urban flowering drive and the lavish Istanbul Tulip Festival. "I promise you two things," Mayor Imamoğlu asserted firmly and loudly: "We will take climate change seriously and under no circumstances open our limited green spaces to development." This last remark fired up the crowd. As the mayor's talk was interrupted by cheers and a long round of applause, there was one burning question begging for an answer: What is it about urban greenery in Istanbul that so markedly energizes politics to the point where even a workshop dedicated to city parks and trees provides the perfect stage for a newly elected mayor to shine before the critical gaze of activists, academics, and urban practitioners?

Urban greening has been a global phenomenon for a while now. Local governments and mayors around the world turn to green spaces to improve urban quality, address climate change, increase the profile of the local real estate market, or simply to boost popularity.[1] Despite being high in demand, green spaces are not uncontroversial, as where and how they are built (or not) may produce socially inequitable outcomes and raise environmental justice concerns. Green spaces do not take up much physical space in contemporary Istanbul,[2] yet they have an increasing weight in politics. Since the early 2010s, not only urban politics but also national fault lines have often pivoted around urban greenery, particularly in conjunction with Istanbul's dwindling urban forests, public parks, community gardens, and waterfront strips. At the heart of Turkey's belated "urban green turn"[3] lies Gezi Park and the wave of protest it instigated in the summer of 2013. As the quintessential urban-environmental mobilization in which an Istanbul public park was both the stage and the cause of the protests, the Gezi uprising has had a lingering impact. The impromptu uprising not only demonstrated the vulnerability of the AKP hegemony but also pointed to counter-hegemonic horizons,[4] including how the emergent urban environmental sensibility gets articulated in politics.[5] While the protests were eventually suppressed, the so-called Gezi spirit has lived on and proliferated across local mobilizations, giving the country's environmental justice activism its character: focused on environmental equity and access yet also motivated by antiauthoritarian sentiments.[6] Since Gezi, occupying, reclaiming, or even caring for urban green spaces have come to signify an alternative sociopolitical horizon antithetical to the existing AKP order, reaching far beyond the social and environmental benefits conventionally attributed to them. New guerilla gardens have popped up in the few remaining vacant plots while existing public parks came under close scrutiny in and beyond Istanbul, often producing ad hoc activism, protests, and occupation against unwanted redevelopment plans.[7]

Growing interest in the mobilizing capacity of the emerging urban-environmental imaginary, however, has not remained exclusive to the opposition. Rather than ignoring the environmental challenge entirely, the governing AKP and its leader, President Erdoğan, chose to absorb it by utilizing a number of "green" strategies. Ranging from greenwashing to co-optation of nature-based solutions, from green gentrification to promoting alternative environmental ethics, the government sought to mold the emergent urban green turn in its own image. Curiously, the government's growing interest in green spaces accompanied, not conflicted with, the party's increasingly authoritarian rule, curtailment of civic rights and liberties, withholding signatures from the Paris Climate Agreement[8] and a deepening economic crisis.[9]

Concerned with this unlikely authoritarian interest in the green agenda, a surprising intersection of two globally salient phenomenon, this article examines Istanbul's urban green spaces as contested sites in the age of authoritarian populism. We focus on Millet Bahçeleri (the Gardens of the Nation), the most ambitious iteration of AKP's urban greening frenzy, in which the central government plans, subsidizes, finances, and builds ornate city parks, bypassing local administrations. Conceived in 2018, the Gardens of the Nation has been used by President Erdoğan as a domestic policy tool in various election campaigns and also promoted abroad as a smart green policy for climate change mitigation and adaptation.[10] By examining how this new breed of gardens has been planned, promoted, discussed, inaugurated, and used, the article questions how and why urban greenery is instrumentalized and co-opted for sociospatial control. By reading these new gardens not simply as green urban interventions but also as political aesthetic enterprises,[11] we evaluate the limits and successes of AKP authoritarianism in undermining an emerging urban-environmental objection and fostering alternative green imaginaries and communities.

In what follows, the article first discusses space-based contentions leading up to the Gezi uprising and the rise of urban-environmental justice activism in and beyond Istanbul. Second, we introduce the Gardens of the Nation initiative as the AKP's major response to rising urban-environmental opposition. The next two sections further examine the initiative and illustrate how the gardens fit into the broader political economy and populist authoritarian rule of the central government, respectively. We also elaborate on how the initiative speaks to, and potentially expands, major directions in critical urban greenery studies. We conclude the article with a discussion on political aesthetics as a means to understand how urban green spaces are used not only to redesign cities but also to divide and regroup the public and rally it behind social and political projects.

The Gezi Moment and the Origins of the Politics of Urban Greenery

Much has been written about the Gezi uprising and its radical openness.[12] As an impromptu objection it was unexpected, nonhierarchical, and without a core leadership to guide it. Gezi unprecedentedly mobilized a highly heterogeneous crowd composed of predominantly first-time activists from very diverse backgrounds. They were on the streets with different motivations, and their expectations from the revolutionary momentum of the uprising did not necessarily converge. Resembling a fairground during the two-week-long occupation by the activists, Gezi Park became a stage where myriad grievances, including rising authoritarianism, the patriarchy, the Kurdish problem, police violence, rampant alcohol taxes, and discrimination against the LGBTQ, as well as the urban and environmental crises, were all voiced.

Despite being the heterotopia that it was, it is still critical to acknowledge what triggered the Gezi uprising to better understand its broad appeal and lingering impact. When a handful of activists rushed to Gezi Park on May 27, 2013, the activists were desperately hoping to stop the construction vehicles driving into the park and uprooting ten to fifteen trees. The vehicle infiltration, according to the activists, was a first step toward clearing the park for a redevelopment project that had been on the table for months. Personally announced and promoted by the then–prime minister Erdoğan himself, the project included pedestrianizing Taksim Square and rebuilding the Ottoman-era military barracks (Topçu Kışlası) that was demolished to clear space for Gezi Park in 1940. Covering most of the park grounds, the new structure was modeled after the barracks and would be utilized as a shopping mall accompanied by luxury condominium units on the upper floors. Planned without any local input and participatory mechanism, the project targeted one of the last remaining green spaces on the European side of the city and would contribute to the ongoing process of commercialization of the public spaces at the city center. Central government's arrogant insistence on designing an urban square, activists' relatable interest in saving a public park from certain demolition, and excessive police brutality found their echo within the larger public. The modest protest that started with a small group of urban activists in Istanbul drew thousands of supporters and grew into a mass antigovernment protest that spread to other cities across the nation.

Despite its spontaneity, the dissent at Gezi was not simply articulated out of thin air. It came into being against the backdrop of what some refer to as bulldozer neoliberalism[13] to indicate the growing weight of construction and extraction sectors in the country's political economy. From the late 2000s onward, the governing AKP has increasingly relied on a growth model that is based on a scheme in which cheap available foreign finance is channeled into construction through public/private real estate development in cities along with energy and mining investments in the countryside.[14] The government saw construction as the engine of the economy as the sector reached an annual growth rate of 9.4 percent between 2010 and 2018, three percentage points greater than the 6.4-percent annual GDP growth rate for the same period.[15] In 2002, a year before AKP came to power, building permits issued by the municipal authorities corresponded to a 15-million-square-meter floor area. This number skyrocketed to 82 million in 2011 and surpassed 200 million in 2016.[16] This aggressive investment in (re)construction targeted first and foremost Istanbul, specifically its last remaining vacant lands, historic quarters, and informally developed neighborhoods with vulnerable title deeds. Green spaces were opened to development, the inner city attracted high-end stores and residents as the well-established working-class neighborhoods were uprooted. While some of this change was done through the hand of the market,[17] urban renewal to a large degree was enabled, or undertaken by, central or local governments, a process scholars have labeled as planned or state-led gentrification.[18]

While high-rises and shopping malls began to dominate the urban fabric in the 2010s, the following decade witnessed the emergence of large-scale urban infrastructure projects as the next stage of construction frenzy. Announced proudly as "mad projects" in conjunction with Erdoğan's 2011 general election campaign, these projects had a heavy carbon footprint and put significant pressure on urban green spaces, wetlands, and forests.[19] Advertised as testaments of modernity, power, and development, the most ambitious of these megaprojects have disproportionally targeted Istanbul: a third bridge on the Bosphorus, a new airport that is supposed to be the largest in the world, and the Istanbul Channel, a massive waterway structure connecting the Black Sea to the Marmara Sea.[20] Thanks to these large-scale undertakings, Turkey began to compete with Brazil and China in terms of total investment in infrastructures and single-handedly absorbed half of global private infrastructure spending in 2015.[21]

Access to land and its development is key for effectively sustaining this model. Legal and institutional frameworks were restructured so that the spatial interventions targeting urban and rural areas can be eased.[22] These measures include a variety of governmental technologies and regulations, including cadastral services, liberalization of land-use regulations, heavy reliance on eminent domain processes, deindustrialization, public-private partnerships, and land reclamation. In Istanbul the pending earthquake threat was used as a pretext for urban renewal schemes, while in the countryside, spatial interventions were justified though a resource independence discourse was used to promote energy, mining, and transportation infrastructures. During the revaluation, gentrification, and commodification of land, many working-class households were pushed out of the city centers,[23] villagers were forced to migrate,[24] and green public spaces were lost. As the construction economy radically transformed the country's urban and rural landscapes, AKP's self-image as the able service provider,[25] as the leader of state-led neoliberal developmentalism,[26] was met with a narrative that is more concerned with environmental ethics and aesthetics of bulldozer neoliberalism and its political ecology. In the late spring of 2013, Gezi Park at Istanbul's Taksim Square became the unexpected stage where competition between these narratives finally became palpable over the destiny of an urban park.

The end of the Gezi events also marks the emergence of two competing and mutually reinforcing environmental practices. As the protestors were expelled from the park by mid-June 2013, they spread across the city and began to hold public forums in different smaller-scale parks of Istanbul. Known as park or neighborhood forums, these modest collectives enabled Gezi veterans to experiment with direct democracy through park assemblies and practices of "commoning" and occupation at the local level.[27] While some of these forums dissolved within a year, those that managed to articulate with local urban-environmental causes established long-lasting networks. Those trained at Gezi now constitute a new breed of activism independent from the established oppositional forces, which often overlook urban and environmental conflicts. The radical openness of the Gezi uprising and park forum experience in its aftermath have contributed to the emergence of new political practices and subjectivities in Turkey.[28] Urban-environmental mobilization, which had remained relatively outside the conventional oppositional forces, have migrated to the center of contentious politics as first-time activists and established political left, too, have gradually come to the movement. Reclaiming the public parks, gardens, and urban forests of Istanbul not only kept the Gezi spirit alive but also provided newcomers an easy access to environmental activism. As such, defending the urban greenery of Istanbul has become a mainstream act of defiance, and simultaneously a symbol of rejection of an intensifying authoritarian urban rule.

In conjunction with rising urban greenery activism, another environmental imaginary and aesthetics was also in the making, calling for alternative environmental aesthetics. In the immediate aftermath of Gezi Park's two-week-long occupation, police barricades were placed around the park, and all the exits and entrances were blocked. The park remained closed until it was ceremoniously reopened three weeks later by the state-appointed governor of Istanbul accompanied by hundreds of police officers. The park was not redone as originally planned. To the contrary, it was renovated and manicured. As all the remnants and memories of the

Gezi commune were cleared from the park, it was turned into the immaculate and lavish garden it had never been. One hundred newly grown trees, 5,000 rose bushes, and 200,000 seasonal flowers were planted, turf was laid for 26 square meters of soil, benches were replaced, a water fountain was erected, the playground was renewed, and the park was expanded by 8,000 square meters toward Taksim Square.[29] As the governor walked along the park and the new park features introduced to the press under heavy police presence, he also reminded all that Gezi Park is for all Istanbulites and said: "Demonstrations hinder the public ability to enjoy it here. . . . People would come to this park with their families. Gezi Park is ready now to embrace all Istanbulites."[30] AKP machinery and Erdoğan were forced to put the Topçu Kışlası project on hold, yet they neither let Gezi be nor withdrew from the politics of urban greenery altogether. Renovation of Gezi Park and its reopening to the public under police surveillance signaled that the green objection could be not only countered but also harnessed to sustain alternative communities.

The Gardens of the Nation: The True Environmentalism

The Gardens of the Nation initiative was first introduced by Recep Tayyip Erdogan himself as a part of his presidential bid leading up to the presidential election of June 24, 2018.[31] In a late-night interview on the state broadcaster TRT, Erdoğan laid out his vision for the presidency and allocated a surprisingly large chunk of time to introduce the new public parks program. Heralded as "the most joyful news" of the night, Erdoğan revealed a rezoning plan that would transform Istanbul's soon-to-be-defunct Atatürk Airport into a massive twelve-square-kilometer public park to be named Millet Bahçesi (the Garden of the Nation). "We want a centrally located garden in the city," Erdoğan explained, and compared the project to its global counterparts: "[They say] that England has great gardens, others do as well. . . . Fine, behold, we will, too" (fig. 1). Admitting the green spaces deficiency in Istanbul, Erdoğan pitched the garden as a much-needed place where "people can easily access with their families and kids, spend great time, eat, drink and roll-over on the grass as they wish."[32]

In the following days, it became clear that Erdoğan's unprecedented monologue on public parks that night was not an accidental digression. The Gardens of the Nation initiative became a major component of Erdoğan's stump speech along the campaign trail and made its mark all the way to the election manifesto of both the party and Erdoğan. It turned out that the project's scope was not limited to a single public park either. There were several Gardens of the Nation projects for Istanbul, and the intention was to spread the gardens to all eighty-one provinces (fig. 2). To accompany the Gardens of the Nation, Erdoğan even developed and

Figure 1. Following President Erdoğan's election promise many renderings flooded the media showcasing the mega urban garden that would replace the defunct Atatürk Airport. This one is by PTMProject, a local landscaping office.

promoted a parallel concept called Millet Kırahathanesi (the Coffeehouse of the Nation), a recreational space where citizens can enjoy government-subsidized coffee, pastries, and internet.[33]

The initiative did not disappear from the agenda even after Erdoğan and his party won the June elections. In fact, Erdoğan made the gardens an important component of the presidential agenda by pledging to complete five, and initiate six, gardens in Istanbul within the first one hundred days in office. The promise was kept. On December 14, 2018, five new public parks totaling roughly 1.5 square kilometers were concomitantly opened in five different neighborhoods of Istanbul in an impressive ceremony starring Erdoğan. In his address, Erdoğan clarified the political work he wants the new gardens to accomplish by stretching the wounds of the Gezi uprising. "Those who wreck and destroy in the name of environmentalism, the perpetrators of Gezi events and those who oppose every single good deed in this country," Erdoğan said in contempt, "should check out these gardens of the nation and see what real environmentalism is."[34]

What counts as true environmentalism? This is a trope President Erdoğan often goes back to in the aftermath of the Gezi uprising.[35] It is noteworthy that, unlike many contemporary authoritarian leaders, Erdoğan refrains from directly attacking or ridiculing the environmental cause. In a nuanced way, and true to the spirit of the post-truth era, he strives to

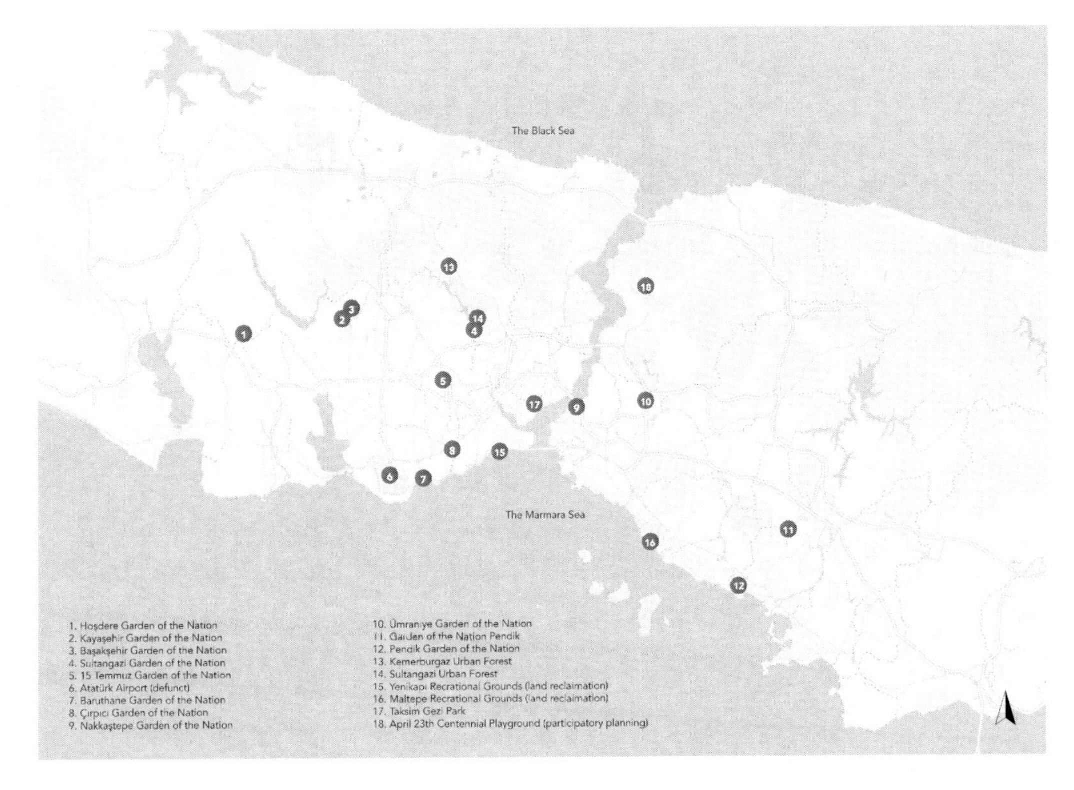

Figure 2. Some urban greening interventions in Istanbul in the aftermath of Gezi uprising in 2013. Map by Deniz Öztürk, Center for Spatial Justice, Istanbul.

reclaim environmentalism with his own definition. In his attempt to define what true environmentalism is, he regularly cites the number of trees his government has planted over the years, brags about the flamboyant roadside landscaping in Istanbul, inaugurates a new national tree planting day, and initiates a zero-waste public program with the first lady, marketing it abroad as a climate change mitigation measure while refusing to sign the Paris Climate Accord.[36] If all these efforts fail to provide him the credibility and recognition he seeks to have, Erdoğan discredits environmentalists but not environmentalism per se. The environmentalist Erdoğan hammers in his speeches is a sham environmentalist who is either a pawn (of foreign powers) or simply alien to local values, folding environmentalism into his culture wars. This culture war not only pits secular environmentalists against conservatives, but also delegitimizes the West as the cradle of environmental ethics, since, as he often reminds his followers, Westerners do not follow the environmental standards that they themselves set.

The name of the gardens initiative fits neatly to this picture. The emphasis on "the nation" (*millet*) echoes "authentic and national" (*yerli ve milli*), the nativist motto that Erdoğan has used for a few years now to draw a sharp contrast between the true owners of the land (including his movement and allies) and the state's external and internal enemies (predominantly political foes and dissidents).[37] While national (*milli*) in this phrase more directly refers to an ethno-religious core, authentic (*yerli*) implies a borderline xenophobic communal exceptionalism with unique and inimitable cultural and attitudinal qualities.[38] Accordingly not only parliamentarians, bureaucrats, and policies but also newspapers, celebrities, and private companies, as well as tastes and habits, are expected to be "authentic and national." An urban greenery project as ambitious as the Gardens of the Nation, then, has to be squarely aligned with the "authentic and national" side of this Manichean worldview, forcing us to imagine green spaces that are not. At this juncture, whether Erdoğan reminds us or not, Gardens of the Nation by definition stands in opposition, first and foremost, to Gezi. The gardens negate what Gezi stood for both in function and aesthetic appeal as we will further elaborate below. To borrow a dichotomy popularized by a pro-Erdoğan youth organization, one could argue that the gardens are not intended to appeal to the *gezici*s (short for a Gezi protester but also literally meaning a traveler or wanderer), but rather to create alternative spaces and environmental imaginaries for *kalıcı*s (permanents, settlers).[39]

Between Safe Greenery and Green Washing

Advertised as new safe urban spaces for families, the gardens also played an important role in AKP's mayoral bid in Istanbul and beyond during the March 31, 2019, local administration elections. All AKP candidates pledged to be part of the initiative and build a garden in the cities that they intended to run. Binali Yıldırım,[40] Erdoğan's pick for Istanbul, on the other hand, promised to increase the number of gardens in the city to thirty-nine, one corresponding to each district.[41] While the March 2019 elections resulted in a major political upset costing AKP major cities, including Istanbul, the Gardens of the Nation initiative has continued to shape the urban fabric and politics. In early 2019, well after a number of gardens were already complete and opened to visitors, the initiative was given legal status and included in the zoning legal framework. Defined as a "large green space that brings the public closer to nature and functions as a disaster evacuation space," the Gardens of the Nation was officially recognized as a new urban green space category next to three already existing categories: the playground, the public park, and the picnic and recreational space. Since then the gardens have a unique sign and color and are included in the zoning plans as such.

Despite having a legal status, there are still a number of aspects of the Gardens of the Nation that make it an ambiguous urban category. It is unclear, for example, what, other than its name, distinguishes such a

Figure 3. Natural gas–powered barbecue stations at Ümraniye Millet Bahçesi (Garden of the Nation), operated by AKP-led Ümraniye District Municipality. Photograph by Sinan Erensü.

garden from a public park. The legislation in fact stipulates a guidebook to be prepared by the Ministry of Environment and Urbanization to standardize details such as size, function, location selection, and tender conditions of these structures. The guideline was published only in May 2020, trailing way behind the initiative itself.[42] Although the gardens have been promoted with delusions of grandeur, they come in different sizes and functions. While Başakşehir Millet Bahçesi, for example, stretches across 465,000 square meters and features an artificial lake, an amphitheater, bicycle lanes, and a Coffee House of the Nation inside the premises, the 59,000-square-meter Baruthane Millet Bahçesi is smaller than many regular public parks in the city.

A much more controversial ambiguity pertains to the choice of garden locations. There exist no criteria regarding the quality of urban land that can be transformed into a Garden of the Nation garden. In some occasions, gardens are developed on top of already existing recreational areas and simply mean a rebranding process undertaken by regulating the unregulated recreational fields. For example, located in the Asian part of the city, the recently opened *Ümraniye* Millet Bahçesi used to be a popular picnic ground and urban forest attracting Istanbulites and their charcoal grills (fig. 3). In 2019, AKP-run District Municipality of Ümraniye decorated the 330,000-square-meter land with recreational furniture, playgrounds, and running and biking tracks. It also updated the restrooms, laid down turf underneath the woods, opened up parking space serviceable to one thousand cars, built a sizable cafeteria, began to monitor the area with security personnel, and called it a Garden of the Nation. In this carefully curated and highly regulated space it is now forbidden to barbecue on personal charcoal grills. Instead, the garden's trademark is five large natural-gas-powered barbecue stations, designed to offer the visitors "the privilege of barbecuing," in the words of the mayor of Ümraniye, "without the usual hassle and smoke."[43] Boosted as "centralized" and "nature-friendly," the stations are overseen by municipal personnel who assist visitors who prefer not to make their own barbecues.[44]

Over-curation and hyper-regulation are not the only reasons why rebranding of green spaces in the form of Gardens of the Nation is problematic. Gardens of the Nation is also used to cover up questionable zoning practices in contemporary Istanbul, to greenwash projects that actually hinder public access to environmental goods. The waterfront Baruthane Millet Bahçesi, located in Bakırköy District's Ataköy neighborhood on the European side of the city, is a case in point. Spanning across eight square kilometers and home to roughly fifty thousand residents, Ataköy is one of the earliest and arguably the most successful middle-class public housing projects in Istanbul, enjoying social amenities and green spaces. When the housing project was first planned in the early 1950s, the two-kilometer-long, two-hundred-meter-wide waterfront strip in front of it was developed into a public beach that remained popular among Istanbulites until 1980s, when the pollution of the Marmara Sea made it unhealthy for swimming. The waterfront was left underfunded and rundown for decades until it was divided into four parcels by the central government and auctioned out to private developers in the early 2010s.

Despite previously being designated exclusively as a tourism zone, the developers were allowed to take advantage of loopholes. Their permit applications for hotel construction were approved only to be later amended to include upscale residential high-rises. After lengthy court proceedings accompanied by Gezi-inspired street demonstrations, the local community, organized under the banner of Bakırköy Urban Defense (Bakırköy Kent Savunması), lost the struggle over the shoreline with the expectation of one relatively small parcel.[45] Squeezed between fourteen ultraluxe con-

Figure 4. Sunday morning at Baruthane Millet Bahçesi (Garden of the Nation), bordering recently completed super-lux waterfront high rises, one of which is occupied by a Hyatt Regency Hotel. Photograph by Sinan Erensü.

dominiums, a Hyatt Regency, a new yacht harbor, and a shopping mall, the smallest parcel's contract was canceled by a court decision thanks to the remnants of a historic gunpowder factory (*baruthane*) and a registered tree found on the premises. In the post-Gezi climate, the government refrained from seeking out a new contract and changed the status of the parcel into a public park. After a few weeks of manicuring, the parcel was inaugurated as Baruthane Millet Bahçesi and proudly promoted as an attempt at increasing people's access to the Marmara Sea (fig. 4).[46]

Sealed off by high-rises on both sides, Baruthane Millet Bahçesi is now one of the leading examples of the Gardens of the Nation initiative as well as an inadvertent reminder of AKP's aggressive capital-friendly urban politics.

Global Rise of Urban Greenery and Authority in the Garden

Interest in urban greenery is not unique to Turkish politics. Urban green spaces, whether in the form of public parks, green belts, or vertical gardens, do have a positive reputation across the world. Many people see urban greenery in a positive lens and enjoy living, working, and commuting in and around it. In response, local governments in the Global North and South alike embrace urban greening projects[47] and other nature-based solutions with the hope of reaping their economic,[48] health-care,[49] ecological, and climate-change-related benefits.[50] Green interventions are not only undertaken at the level of city government, they are also discursively and financially supported as state-of-the-art policies by international institutions such as the World Bank, various United Nations (UN) agencies, and the European Union.

The fact that they are among the most desired urban amenities, however, does not mean they are immune to politics or necessarily produce socially and ecologically equitable outcomes.[51] In fact, despite being promoted as benign, common sense, smart, and nature-based solutions to various ills of urbanization, urban greening projects are increasingly being contested by the local communities and scrutinized by scholars.[52] Green investments are not equally distributed across cities and often disproportionately favor the already affluent parts of the cities, or they are steered by urban growth alliances to benefit future real estate developments and/or commercial interest. Because far too often, green interventions are planned and implemented in a top-down fashion without much attention to neighborhood histories and demographic and social dynamics. Urban researchers and political ecologists have also been arguing that green planning orthodoxy cannot help but be an extension of larger neoliberal forces, sometimes despite the best intentions.[53] Greening spaces and projects help local authorities to brand cities as green to attract capital. Another well-documented aspect of neoliberal urban greening is what scholars term "green gentrification" (or eco-gentrification), in which expansion or improvement of green amenities inflates the nearby real estate market, harming lower income communities.[54] Green gentrification is an exclusionary process and often means dispossession and displacement for the urban poor and ethnic and racial minorities.

The Gardens of the Nation, too, suffer from most of the shortcomings of urban greening interventions highlighted in the literature. The new gardens are imposed, rather than collectively designed through local input. Although they are promoted for their disaster risk management capacities, the central government refuses to take responsibility against earthquake and climate change threats that loom large over the city.[55] The processes behind where the new gardens are to be located are not transparent. They are promoted as increasing citizen access to urban environments, yet some of the gardens are simply being built over already existing, yet underplanned and inadequately designed recreational city grounds. Others, unabashedly, are built next to highly controversial high-end waterfront condominiums for greenwashing purposes, therefore approving, not challenging, neoliberal urbanism. While we lack detailed research on how the new gardens impact lower income neighborhoods, the government-affiliated press already proudly announces rent increases in neighborhoods expecting the new gardens nearby.[56]

Despite all these characteristics that new city gardens of Turkey share in common with their urban greening counterparts across the globe, we believe that they also retain some unique aspects that would enhance our understanding of how urban green is folded into politics. Most critically, unlike most urban greenery interventions, the Gardens of the Nation is not about local initiatives; the gardens are personally promoted and overseen by President Erdoğan himself. By this token, they symbolize and contribute to the centralization of power in the country and help Erdoğan undercut the reach of municipal governments

While, for example, the controversy surrounding the Baruthane Garden above speaks to the assumed—if not necessarily the actual—role of green spaces in obscuring urban disputes, it also exemplifies how the central government expands its authority over local politics through the seemingly mundane politics of urban greenery. Thus far, we have illustrated that President Erdoğan is not only the originator and the patron of the Gardens of the Nation initiative; he also personally attends most inauguration ceremonies and skillfully turns every garden visit into a political rally. It is critical to note, however, that the authority performed over local politics through the Gardens of the Nation is not just discursive but also political-economical.

The responsibility to run the gardens initiative is given by President Erdoğan to the Ministry of Environment and Urbanization. Under the ministry, TOKI (Mass Housing Agency of Turkey), the infamous public developer best known for its heavy-handed urban renewal program resulting in evictions and gentrification, is tasked with the responsibility to design, finance, and build the gardens.[57] As of May 2020, there are eleven gardens completed by TOKI, while the agency has nineteen projects under construction, twenty-one ready for tender request, and 103 at an early stage of planning.[58] Planned behind closed doors with no local input, the gardens cause fluctuations in the local real estate market, too, by creating advantage for those with prior knowledge.[59] Another dimension of this secrecy and closeness pertains to the planning of the parks. One of the noticeable actors in the project development is ON Tasarım, an urban design/landscape architecture company based in Ankara and Istanbul, infamously known for the design of Taksim Square that triggered the uprising in 2013. The company had been commissioned to design thirteen gardens so far, including Istanbul Millet Bahçesi, the largest project of its kind, which is planned to repurpose Atatürk Airport.

Gardens of the Nation are not cheap undertakings either. While the eventual costs depend on the size and location of the garden, Pendik Millet Bahçesi, located in Istanbul on more than 250,000 square meters of land, for example, is expected to cost as much as 66 million Turkish lira (roughly $8.8 million).[60] An ongoing garden project in Ankara, highly controversial because the contractor was handpicked by TOKI instead of selected through a tender, is reported to have a colossal 398.5 million Turkish lira (roughly $52 million) price tag on it.[61] These are significant sums for the city and district municipalities as most of them are heavily indebted.[62] Yet thankfully, all the costs of the initiative are covered by Ankara, enabling the central government to be both politically and financially active in local politics. This is a much-needed political reach given AKP's diminishing electoral success in metropoles. The logos of TOKI as well as the Ministry of Environment and Urbanization are clearly marked at the gates, giving the central administration the bragging rights and the publicity opportunity. Yet, many gardens fall under the jurisdiction of municipalities, triggering potential conflicts between the central and local government especially if the latter is run by the opposition. To overcome such conflicts, Ankara either ignores the metropolitan municipality and finds a friendly district municipality to cooperate with or tries hard to change the status of the land in question in a top-down fashion.

Closing Remarks: Green Aesthetics and Community

Despite problems in the top-down planning and implementation, it would be inaccurate to suggest that the Gardens of the Nation are unwanted urban amenities. Particularly in Istanbul, where urban residents are desperate for recreational grounds to breathe and relax, there is little reason why the new gardens would not be as popular as any other green spaces. Therefore it is not surprising that the Gardens of the Nation in Istanbul, most of which used to be under-designed recreational grounds anyway, are frequented by Istanbulites. The purpose of this article is not to pass a value judgment against the new government gardens. To the contrary, what we aim to accomplish in this article is to illustrate the critical role that they have played for President Erdoğan to counter a formidable environmental challenge that became palpable with the Gezi protests.

We contend that beyond being a simple policy maneuver, the case of the Gardens of the Nation also speaks to a growing concern regarding how populist authoritarianism relates to the environment. The success of populist authoritarianism that we encounter in different shapes and forms around the globe lies in its ability to speak "in the name of the people" while, at the same time, setting the standards for "who the people are" in a pretty racialized and nativist manner. Environmental matters are highly relevant to these definitions, as they often bring about questions of land, culture, and taste.[63] Therefore, it is no coincidence that most contemporary populist authoritarian leaders take strong conservative positions in environmental disputes. What characterizes these positions is the belief that people's sovereignty over the native land is at odds with, and superior to, universally recognized environmental principles and environmental justice concerns. Erdoğan's interest in urban greenery, we argue, illustrates how populist leaders maintain authority and command not only by suppressing the environmental critiques, but also by fostering alternative environmental tastes, ethics, aesthetics.

Positioned in direct opposition to the urban-environmental challenge of Gezi, the gardens not only help the government to take the urban space back but also exemplify what "true environmentalism" loyal to the blood and soil may look like. "True environmentalism" is not a textbook definition to be learnt; it is also aesthetically experienced.[64] Senses play an important role in establishing the relationship between people and politics. What we experience with our senses (e.g., "the beautiful") is a subjective judgment but at the same time, it seeks others' recognition and acknowledgment. The moment of enunciation of an aesthetic term signifies the immediate transition from a subjective judgment to the commonsensical, which is bound to fail because of the impossibility of consensus on sensory experiences. Hence, to name something "beautiful" as an aesthetic judgment necessarily leads to two different camps but always maintains the futile search for unity. The new government gardens, likewise, speak to our senses; they foster a certain environmental taste and build communities that appreciate that taste. Istanbul-based urban activists and landscape architects that we talked to find the gardens kitschy, ostentatious, and overdesigned. For many others, including but not exclusively the staunch AKP supporters, they are clean, safe, and perfect for family weekend outings and Instagram photoshoots.[65]

In this respect, the Gardens of the Nation initiative that was presented by Erdoğan as an "authentic and national" (*yerli and milli*) project aiming to re-create the beauties of Istanbul and beyond for the use of the nation functions as a dividing force in the politics of urban greenery. *The Gardens of the Nation Guidebook* (*Millet Bahçeleri Rehberi*), which was released by the Ministry of Environment and Urbanization as we conclude the second draft of this article, provides an illuminating perspective into this political aesthetic project. The booklet describes its purpose as to "set a common language . . . around the Gardens of the Nation that would reflect the nation's welfare and dynamism and symbolize national image and identity."[66] While listing some very nonspecific and nonbinding guidelines as to how to plan, design, and build the new gardens, the booklet most remarkably traces a Muslim-Turkish tradition with regard to parks and gardens. Through this detailed historic account, the guidebook establishes the authenticity of the Gardens of the Nation initiative and categorize it as the latest iteration of a proud gardening lineage with deep roots in Seljuk, Ottoman, and early Republican eras.

Evident in the content and tone of the Green Spaces Workshop that we discussed in the introduction, the newly elected mayor of Istanbul, too, cannot escape the politics of urban greenery. The IMM administration under Mayor Imamoğlu strives to distance and differentiate itself from the Erdoğan government's heavy-handed handling of urban environment matters. After all, one of the major criticisms of urban opposition in Turkey in the past decade has been the lack of participation in urban governance, and the environmental legacy of the Gezi protests still motivates and inspires the opposition. The new municipal cadres now include Gezi veterans alongside experts, activists, and academics who have been active in urban-environmental opposition in Istanbul for the last two decades. Around the time the central government has been introducing new gardens in a top-down fashion, Istanbul's newly elected local government was holding consultative workshops on topics ranging from earthquake preparedness to water management, from urban farming to urban greenery. One of the boldest moves of Major Imamoğlu has been his opposition to President Erdoğan's latest environmentally controversial megaproject: the new waterway connecting the Black Sea to the Marmara Sea. Succinctly summarized with the slogan "it is either Istanbul or the channel," the major has mobilized IMM against the project. To better document and protect Istanbul's fragile ecosystem, a new directorate with the title Urban Ecological Services has been defined under IMM. Last but not least, two new municipal recreational areas have been inaugurated in the last remaining urban forests of the city and promoted as minimally designed, in contrast with the Gardens of the Nation. While these developments are most welcome from an environmental standpoint, the extent to which they will help overcome the authoritarian polarization over urban greenery is yet to be seen.

Sinan Erensü is an assistant professor of sociology at Boğaziçi University, Istanbul. He is an urban political ecologist working on environmental disputes, energy democracy, and speculative urbanism. He previously taught at the University of Minnesota and Northwestern University.

Barış İne is an independent researcher affiliated with the Center for Spatial Justice, a nonprofit located in Istanbul. He holds an MA in political science from the University of Minnesota.

Yaşar Adanalı is a PhD candidate at the Technical University of Berlin working on urban transformation and urban grassroots movements. He is one of the founding members of the Center for Spatial Justice and its previous director. He has been teaching participatory planning and design courses in Mundus Urbano Masters Programme at the Technical University of Darmstadt since 2009.

Notes

Sinan Erensü acknowledges research funding support from the European Research Council (ERC) under the European Union's Horizon 2020 research and innovation program (grant agreement No. 680313).

1. Anguelovski et al., "New Scholarly Pathways"; Gould and Lewis, *Green Gentrification*; Wolch et al., "Urban Green Space."

2. Having only 2.2 percent of its land allocated for parks or gardens, Istanbul ranks at the bottom of a forty-city list compiled by the World Cities Culture Forum. Similarly, Green Cities Index published by TravelBird, a Dutch travel agency, ranks Istanbul second to last in a fifty-city list (Suliman, "Green City Getaways").

3. By "urban green turn" we refer to the proliferation of urban greening initiatives and the popularization of nature-based solutions to remedy social and environmental ills of the cities (Anguelovski et al., "New Scholarly Pathways").

4. Bakıner, "'Spirit of Gezi'"; Konya, "Breaking Billboards"; Çınar, "Negotiating the Foundations."

5. Akbulut, "A Few Trees"; Arsel et al., "A Few Environmentalists?"; Erensü and Karaman, "Work of a Few Trees"; Mert, "The Trees in Gezi Park"; Özdüzen, "Spaces of Hope"; Özkaynak et al., "The Gezi Park Resistance."

6. Pellow, "Critical Environmental Justice."

7. Urban green spaces that have witnessed major antiredevelopment activism in the aftermath of Gezi include Validebağ Woods, Roma Bostan, İlya's Bostan, Albatros Park, and Kuruçeşme Park in Istanbul; Middle East Technical University (METU) Forest and 100. Yıl Park Ankara; Diyarbakır City Forest, Karayolları Park, and Hevsel Gardens in Diyarbakır; Bankalar Park in Amasya; and Mehmet Bozgeyik Park in Gaziantep.

8. Turhan et al., "Beyond Special Circumstances"; Cerit Mazlum, "Still Playing Alone."

9. Akçay and Gürgen, "The Making."

10. Yavuz, "İklim Zirvesinde."

11. Rancière, *The Politics of Aesthetics*.

12. Açıksöz and Korkman, "Masculinized Power, Queered Resistance"; Ay and Miraftab, "Invented Spaces"; Diken, "The Emancipated City"; Erensü and Karaman, "The Work of a Few Trees"; Evren, "On the Joy"; Ertür, "Barricades"; Gambetti, "Occupy Gezi"; Karakayalı and Yaka, "The Spirit of Gezi"; Tuğal, "Resistance Everywhere"; Turhan, "'Democracy Happens'"; Yıldız, "Cruising Politics"; Yörük, "The Long Summer"; Zengin, "What is Queer About Gezi?"

13. Lovering and Türkmen, "Bulldozer Neoliberalism"; see also Eder, "Türk Usulü."

14. Adaman et al., "Hitting the Wall."

15. World Bank, "GDP Growth Turkey."

16. TURKSTAT, "Yapı İzin İstatistikleri."

17. Yetişkul and Demirel, "Assembling Gentrification"; Özdemir and Selçuk, "From Pedestrianisation."

18. İslam, "Current Urban Discourse"; İslam and Sakızoğlu, "State-Led Gentrification"; Kocabaş and Gibson, "Planned Gentrification."

19. The Mega Istanbul research team defines what counts as a mega project and lists and visualizes more than 120 ambitious real estate and infrastructure projects that Istanbul showcases ("MegaIstanbul").

20. The area reserved for the latter two corresponds to 6 percent of Istanbul's surface area.

21. *Hürriyet Daily News*, "Turkey Absorbs Almost Half."

22. Atasoy, "Repossession."

23. Kuyucu and Ünsal, "Urban Transformation."

24. Adaman et al., "Neoliberal Developmentalism"; Erensü, "Turkey's Hydropower"; Evren, "The Rise and Decline"; Öztürk, Jongerden, and Hilton, "Commodification and the Social Commons."

25. Adanalı, "The Park Revolution."

26. Adaman and Akbulut, "Erdoğan's Three-Pillared Neoliberalism."

27. Akçalı, "Popular Assemblies"; İnceoğlu, "The Gezi Resistance"; Pelivan, "Going Beyond."

28. Karakayalı and Yaka, "The Spirit of Gezi."

29. *Anadolu Ajansı,* "Gezi Parkı 8."

30. *Hürriyet*, "Gezi Parkı Açıldı."

31. Paker, "The Politics of Serving."

32. *Hürriyet*, "Erdoğan."

33. *Euronews*, "Cumhurbaşkanı Erdoğan'dan 'Millet Kıraathaneleri.'"

34. *Diken*, "Gezi'yi Yapanlar."

35. Adaman and Akbulut, "Erdoğan's Three-Pillared Neoliberalism."

36. Turhan et al., "Beyond Special Circumstances."

37. Others draw parallels between the contemporary Gardens of the Nation and its namesake public parks of late Ottoman-, early Republican-era modernism (Duru, "Türkiye'ye Özgü bir Proje"; Memlük, "Osmanlı Modernleşmesi ile").

38. Bora, "Zamanın Kelimeleri," 192–206.

39. Küçük and Türkmen, "Remaking the Public," discusses at length the slogan *gezici değil, kalıcı gençlik* (permanent youth, not wanderers) in the context of the new nationalist cosmology in Turkey.

40. Considered to be Erdogan's right hand, Binali Yıldırım proclaims to be the man behind the megaprojects with his fourteen-year-long role as the minister of transport.

41. Although Yıldırım lost the race, the central government has been pursuing this project as a means to compete with the new IMM administration.

42. Ministry of Environment and Urbanization, "Millet Bahçeleri Rehberi."

43. Aksu, "Millet Bahçesine Merkezi Mangal."

44. Aksu, "Millet Bahçesine Merkezi Mangal."

45. In one of several demonstrations, about a thousand demonstrators from the local community marched from Bakırköy Town Square to the Ataköy coast chanting, "We will take back the Ataköy shore." Demonstrators demanded the suspension of construction until the legal process is finalized (*Habertürk*, "Ataköy sahilinde").

46. Levent, "Ataköy Sahili Halka Teslim."

47. In fact, IMM's interest in urban greenery precedes the gardens initiative. Founded in 1997 during Erdoğan's first mayoral term, Istanbul Tree and Landscape Corporation (Ağaç AŞ) is one of the oldest municipal enterprises, running on a hefty budget of eight hundred million Turkish lira as of 2018 (BBC Turkish, "İstanbul Büyükşehir Belediyesi").

48. Heckert and Mennis, "The Economic Impact."

49. Lee et al., "Value of Urban."

50. Wolch et al., "Urban Green Space."

51. Anguelovski et al., "New Scholarly Pathways."

52. Pearsall and Anguelovski, "Contesting and Resisting"; Kotsila, "Nature Based-Solutions"; Anguelovski et al., "New Scholarly Pathways."

53. Anguelovski et al., "From Landscapes of Utopia"; Anguelovski et al.,

"New Scholarly Pathways"; Kotsila, "Nature Based-Solutions"; Brand, "Green Subjection."

54. Gould and Lewis, *Green Gentrification*; Dooling, "Ecological Gentrification"; Quastel, "Political Ecologies."

55. Turkey remains one the few countries that has not ratified the Paris Climate Agreement. Istanbul has long been expecting a major earthquake of around 7.5 magnitude, yet the number of buildings that are not expected to withstand it is around fifty thousand by conservative estimates.

56. *Yeni Şafak*, "Erdoğan'ın açıklamasıyla."

57. Çavuşoğlu and Julia Strutz, "Producing Force and Consent"; Karaman, "Urban Renewal in Istanbul."

58. Turapoğlu, "TOKİ 81."

59. Tabak, "Millet Bahçesi'nin Adı Fiyatları Uçurdu."

60. Güvemli, "Türkiye'nin Milyonluk Bahçeleri."

61. Toker, "AKM Millet Bahçesinde Neler Oluyor?"

62. *Sözcü*, "AKP'li Belediyeler Borç Dağı Bıraktı."

63. McCarthy, "Authoritarianism," 302.

64. Here, we follow a Rancierian approach to the relationship between politics and aesthetics (Rancière, *The Politics of Aesthetics*). Exclusionary and inclusionary dynamics of spatial aesthetics can be found in other urban conflicts. Ruez, in his work on Park 51 in New York, for example, underlines the ways in which the debate over the location of an Islamic community center and a mosque demonstrates the limits of public discussion around the existing distribution of the sensible with regard to Islamophobia, citizenship, and community ("Partitioning the Sensible"). Ghertner, on the other hand, applies a similar framework to the city of Delhi to explain how the whole city has been transformed into an aesthetic project pursuing world-class standards (*Rule by Aesthetics*).

65. On conservative and gendered expectations from public parks, see Alkan Zeybek, "Bir Aile Mekanında."

66. Ministry of Environment and Urbanization, "Millet Bahçeleri Rehberi," 10–11.

References

Açıksöz, Salih Can, and Zeynep Korkman. "Masculinized Power, Queered Resistance." Society for Cultural Anthropology, October 31, 2013. https://culanth.org/fieldsights/masculinized-power-queered-resistance.

Adaman, Fikret, and Bengi Akbulut. "Erdoğan's Three-Pillared Neoliberalism: Authoritarianism, Populism, and Developmentalism." *Geoforum* 124 (2021): 279–89.

Adaman, Fikret. Bengi Akbulut, Yahya Madra, and Şevket Pamuk. "Hitting the Wall: Erdoğan's Construction-Based, Finance-Led Growth Regime." *Middle East in London* 10, no. 3 (2014): 7–8.

Adaman, Fikret, Murat Arsel, and Bengi Akbulut. "Neoliberal Developmentalism, Authoritarian Populism, and Extractivism in the Countryside: The Soma Mining Disaster in Turkey." *Journal of Peasant Studies* 46, no. 3 (2019): 514–36.

Adanalı, Yaşar Adnan. "The Park Revolution" *Topos Magazine*: *The International Review of Landscape Architecture and Urban Design*, no. 85 (2013): 46–51.

Akbulut, Bengi. "'A Few Trees' in Gezi Park: Resisting the Spatial Politics of Neoliberalism in Turkey." In *Urban Forests, Trees, and Greenspace: A Political Ecol-*

ogy Perspective, edited by Anders Sandberg, Adrina Bardekjian, and Sadia Butt, 245–59. New York: Earthscan, 2014.

Akçalı, Emre. "Do Popular Assemblies Contribute to Genuine Political Change? Lessons from the Park Forums in Istanbul." *South European Society and Politics* 23, no. 3 (2018): 323–40.

Akcay, Ümit, and Ali Riza Güngen. "The Making of Turkey's 2018–2019 Economic Crisis." Hochschule für Wirtschaft und Recht Berlin, Institute for International Political Economy (IPE), working paper no. 120/2019.

Aksu, Fatma. "Millet Bahçesine Merkezi Mangal." *Hürriyet*, August 15, 2019. https://www.hurriyet.com.tr/gundem/millet-bahcesine-merkezi-mangal-41302210.

Alkan Zeybek, Hilal. "Bir Aile Mekanında Cinsiyet, Cinsellik ve Güvenlik." In *Neoliberalizm ve Mahremiyet: Türkiye'de Beden, Sağlık ve Cinsellik*, edited by Cenk Özbay, Ayşecan Terzioğlu, and Yeşim Yasin, 227-243. Istanbul: Metis, 2011.

Anadolu Ajansı. "Gezi Parkı 8 bin Metrekare Büyüyor." July 2, 2013. https://www.aa.com.tr/tr/turkiye/gezi-parki-8-bin-metrekare-buyuyor/234893.

Anguelovski, Isabelle, Anna Livia Brand, James J. T. Connolly, Esteve Corbera, Panagiota Kotsila, Justin Steil, Melissa Garcia-Lamarca, Margarita Triguero-Mas, Helen Cole, Francesc Baró, Johannes Langemeyer, Carmen Pérez del Pulgar, Galia Shokry, Filka Sekulova, and Lucia Argüelles Ramos. "Expanding the Boundaries of Justice in Urban Greening Scholarship: Toward an Emancipatory, Antisubordination, Intersectional, and Relational Approach." *Annals of the American Association of Geographers* 6 (2020): 1–27.

Anguelovski, Isabelle, James Connolly, and Anna Livia Brand. "From Landscapes of Utopia to the Margins of the Green Urban Life: For Whom Is the New Green City?" *City* 22, no. 3 (2018): 417–36.

Anguelovski, Isabelle, James J. T. Connolly, Melissa Garcia-Lamarca, Helen Cole, and Hamil Pearsall. "New Scholarly Pathways on Green Gentrification: What Does the Urban 'Green Turn' Mean and Where Is It Going? *Progress in Human Geography* 43, no.6 (2019): 1064–86.

Arsel, Murat, Firket Adaman, and Bengi Akbulut. "A Few Environmentalists? Interrogating the 'Political' in the Gezi Park. In *Neoliberal Turkey and Its Discontents: Economic Policy and the Environment under Erdogan*, edited by Fikret Adaman, Bengi Akbulut, and Murat Arsel, 191–206. London: I. B. Tauris, 2017.

Atasoy, Yıldız. "Repossession, Re-Informalization, and Dispossession: The 'Muddy Terrain of Land Commodification in Turkey." *Journal of Agrarian Change* 17, no. 4 (2017) 657–79.

Ay, Deniz, and Faranak Miraftab. "Invented Spaces of Activism: Gezi Park and Performative Practices of Citizenship." In *The Palgrave Handbook of International Development*, edited by Jean Grugel and Daniel Hammett, 555–74. London: Palgrave Macmillan, 2016.

Bakıner, Onur. "Can the 'Spirit of Gezi' Transform Progressive Politics in Turkey?" In *The Making of a Protest Movement in Turkey*, edited by Umut Özkırımlı, 103–120. London: Palgrave, 2014.

BBC Turkish. "İstanbul Büyükşehir Belediyesi: Türkiye'nin en Büyük Yerel Yönetimi." May 8, 2019. https://www.bbc.com/turkce/haberler-turkiye-48180564.

Bora, Tanıl. *Zamanın Kelimeleri: Yeni Türkiye'nin Siyaset Dili*. Istanbul: İletişim, 2018.

Brand, Peter. "Green Subjection: The Politics of Neoliberal Urban Environmental Management." *International Journal of Urban and Regional Research* 31, no. 3 (2007): 616–32.

Çavuşoğlu, Erbatur, and Julia Strutz. "Producing Force and Consent: Urban Transformation and Corporatism in Turkey." *City* 18, no. 2 (2014): 134–48.

Cerit Mazlum, Semra. "Turkey and Post-Paris Climate Change Politics: Still Playing Alone." *New Perspectives on Turkey* 56 (2017): 145–52.

Checker, Melissa. "Wiped Out by the 'Greenwave': Environmental Gentrification and the Paradoxical Politics of Urban Sustainability." *City and Society* 23, no. 2 (2011): 210–29.

Çınar, Alev. "Negotiating the Foundations of the Modern State: The Emasculated Citizen and the Call for a Post-patriarchal State at Gezi Protests." *Theory and Society* 48, no. 3 (2011): 453–82.

Dikeç, Mustafa. "Space, Politics, and the Political." *Environment and Planning D: Society and Space* 23, no. 2 (2005): 171–88.

Diken, Bülent. "The Emancipated City: Notes on Gezi Revolts." *Journal for Cultural Research* 18, no. 4 (2014): 315–28.

Dooling, Sarah. "Ecological Gentrification: A Research Agenda Exploring Justice in the City. *International Journal of Urban and Regional Research* 33, no. 3 (2009): 621–39.

Duru, Bülent. "Türkiye'ye Özgü bir Projenin Görünmeyen Yüzü: Millet Bahçeleri." *Mimarlık*, no. 412 (2020): 16–19.

Eder, Mine. "Türk Usulü Buldozer Neoliberalleşmeyi Anlamak: AKP'nin Politik Ekonomisi ve Ötesi." In *Türkiye'de Yeni* İktidar *Yeni Direniş: Sermaye-Ulus-Devlet Karşısında Ulus* Ötesi *Müşterekler*, edited by Yahya Madra, 47–56. Istanbul: Metis, 2015.

Erensü, Sinan. "Turkey's Hydropower Renaissance: Nature, Neoliberalism, and Development in the Cracks of Infrastructures." In *Neoliberal Turkey and its Discontents: Economic Policy and the Environment under Erdogan*, edited by Fikret Adaman, Bengi Akbulut, and Murat Arsel, 120–46. London: I. B. Tauris, 2017.

Erensü, Sinan, and Ozan Karaman. "The Work of a Few Trees: Gezi, Politics, and Space." *International Journal of Urban and Regional Research* 41, no. 1 (2017): 19–36

Ertür, Başak. "Barricades: Resources and Residues of Resistance." In *Vulnerability in Resistance*, edited by Judith Butler, Zeynep Gambetti, and Leticia Sabsay, 97–122. Durham, NC: Duke University Press, 2016.

Euronews. "Cumhurbaşkanı Erdoğan'dan 'Millet Kıraathaneleri' Projesi." June 7, 2018. https://tr.euronews.com/2018/06/07/erdogan-dan-millet-kiraathaneleri-projesi.

Evren, Erdem. "On the Joy and Melancholy of Politics." *Cultural Anthropology: Sightings*, October 31, 2013. https://culanth.org/fieldsights/on-the-joy-and-melancholy-of-politics.

Evren, Erdem. "The Rise and Decline of an Anti-Dam Campaign: Yusufeli Dam Project and the Temporal Politics of Development. *Water History* 6, no. 4 (2014): 405–19.

Gambetti, Zeynep. "Occupy Gezi as Politics of the Body." In *The Making of a Protest Movement in Turkey*, edited by Umut Özkırımlı, 103–120. London: Palgrave, 2014.

Ghertner, D. Asher. *Rule by Aesthetics: World-Class City Making in Delhi*. Oxford: Oxford University Press, 2015.

Goonewardena, Kanishka. "The Urban Sensorium: Space, Ideology, and the Aestheticization of Politics. *Antipode* 37, no. 1 (2005): 46–71.

Gould, Kenneth A., and Tammy L. Lewis. *Green Gentrification: Urban Sustainability and the Struggle for Environmental Justice*. New York: Routledge, 2016.

Güvemli, Özlem. "Türkiye'nin Milyonluk Millet Bahçeleri." *Sözcü*, March 30, 2019. https://www.sozcu.com.tr/2019/gundem/turkiyenin-milyonluk-millet-bahceleri-4225094/.

Habertürk. "Ataköy sahilinde 1000 kişilik eylem!" September 20, 2014. https://www.haberturk.com/gundem/haber/992215-atakoy-sahilinde-1000-kisilik-eylem.

Hammond, Timur, and Elizabeth Angell. "Is Everywhere Taksim? Public Space and Possible Publics. *Jadalliya*, June 9, 2013. https://www.jadaliyya.com/Details/28755.

Heckert, Megan, and Jeremy Mennis. "The Economic Impact of Greening Urban Vacant Land: A Spatial Difference-in-Differences Analysis." *Environment and Planning A: Economy and Space* 44, no. 12 (2012): 3010–27.

Hürriyet. "Gezi Parkı Açıldı." July 8, 2013. https://www.hurriyet.com.tr/gundem/gezi-parki-acildi-23676906.

Hürriyet. "Turkey Absorbs Almost Half of Global Private Infrastructure Investment in 2015." June 14, 2016. https://www.hurriyetdailynews.com/turkey-absorbs-almost-half-of-global-private-infrastructure-investment-in-2015-world-bank-100463.

Hürriyet. "Erdoğan: Atatürk Havalimanı millet bahçesi olacak." May 23, 2018. https://www.hurriyet.com.tr/gundem/erdogan-ataturk-havalimanini-millet-bahcesi-olacak-40846401.

Hürriyet Daily News. "Turkey Absorbs Almost Half of Global Private Infrastructure Investments in 2015." June 14, 2016. https://www.hurriyetdailynews.com/turkey-absorbs-almost-half-of-global-private-infrastructure-investment-in-2015-world-bank-100463.

İnceoglu, Irem. "The Gezi Resistance and Its Aftermath: A Radical Democratic Opportunity?" *Soundings*, no. 57 (2014): 23–34.

İslam, Tolga. "Current Urban Discourse, Urban Transformation, and Gentrification in Istanbul." *Architectural Design* 80, no. 1 (2010): 58–63.

İslam, Tolga, and Bahar Sakızlıoğlu. "The Making of, and Resistance to, State-Led Gentrification in Istanbul, Turkey." In *Global Gentrifications: Uneven Development and Displacement*, edited by Loretta Lees, Hyun Bang Shin, and Ernesto López-Morales, 245–64. Bristol, UK: Bristol University Press, 2015.

Karakayalı, Serhat, and Özge Yaka. "The Spirit of Gezi: The Recomposition of Political Subjectivities in Turkey." *New Formations*, no. 83 (2014): 117–38.

Karaman, Ozan. "Urban Renewal in Istanbul: Reconfigured Spaces, Robotic Lives." *International Journal of Urban and Regional Research* 37, no. 2 (2013): 715–33.

Kocabas, Arzu, and Michael S. Gibson. "Planned Gentrification in Istanbul: The Sulukule Renewal Area 2005–2010." *International Journal of Sustainable Development and Planning* 6, no. 4 (2011): 420–46.

Konya, Nazlı. "Breaking Billboards: Protest and a Politics of Play." *Contemporary Political Theory* 20, no. 2 (2021): 250–71.

Kotsila, Panagiota, Isabelle Anguelovski, Francesc Baró, Johannes Langemeyer, Filka Sekulova, and James J. Connolly. "Nature-based Solutions as Discursive Tools and Contested Practices in Urban Nature's Neoliberalisation Processes." *Environment and Planning E: Nature and Space* 4, no. 2 (2021): 252–74.

Küçük, Bülent, and Ceren Özselçuk. "Fragments of the Emerging Regime in Turkey: Limits of Knowledge, Transgression of Law, and Failed Imaginaries." *South Atlantic Quarterly* 118, no. 1 (2019): 1–21.

Küçük, Bülent, and Buket Türkmen. "Remaking the Public through the Square: Invention of the New National Cosmology in Turkey." *British Journal of Middle Eastern Studies* 47, no.2 (2020): 247–63.

Kuyucu, Tuna, and Özlem Ünsal. "Urban Transformation as State-Led Property Transfer: An Analysis of Two Cases of Urban Renewal in Istanbul." *Urban Studies* 47, no. 7 (2010): 1479–99.

Lee, Andrew Chee Keng, Hannah C. Jordan, and Jason Horsley. "Value of Urban Green Spaces in Promoting Healthy Living and Wellbeing: Prospects for Planning." *Risk Management and Healthcare Policy* 8 (2015): 131–37.

Levent, Sefer. "Ataköy Sahili Halka Teslim." *Hürriyet*, April 1, 2018. https://www.hurriyet.com.tr/yazarlar/sefer-levent/atakoy-sahili-halka-teslim-40790538.

Lovering, John, and Hade Türkmen. "Bulldozer Neo-Liberalism in Istanbul: The State-Led Construction of Property Markets, and the Displacement of the Urban Poor." *International Planning Studies* 16, no. 1 (2011): 73–96.

McCarthy, James. "Authoritarianism, Populism, and the Environment: Comparative Experiences, Insights, and Perspectives." *Annals of the American Association of Geographers* 109, no. 2 (2019): 301–13.

MegaIstanbul. https://en.megaprojeleristanbul.com (accessed May 31, 2021).

Memlük, Yalçın. "Osmanlı Modernleşmesi ile Ortaya Çıkan Bir Kentsel Mekan Olarak Millet Bahçeleri." *Türkiye Sağlıklı Kentler Birliği*, August 16, 2017. https://www.skb.gov.tr/osmanli-modernlesmesi-ile-ortaya-cikan-bir-kentsel-mekan-olarak-millet-bahceleri-s25212k/.

Mert, Ayşem. "The Trees in Gezi Park: Environmental Policy as the Focus of Democratic Protests." *Journal of Environmental Policy and Planning* 21, no. 5 (2019): 593–607.

Ministry of Environment and Urbanization. "Millet Bahçeleri Rehberi." Online guidebook. https://webdosya.csb.gov.tr/db/mpgm/editordosya/milletbahcesirehber.pdf (accessed May 31, 2021).

Özdemir, Dilek, and İrem Selçuk. "From Pedestrianisation to Commercial Gentrification: The Case of Kadıköy in Istanbul." *Cities* 65 (2017): 10–23.

Özdüzen, Özge. "Spaces of Hope in Authoritarian Turkey: Istanbul's Interconnected Geographies of Post-Occupy Activism." *Political Geography* 70 (2019): 34–43.

Özkaynak, Begüm, Cem İskender Aydın, Pınar Ertör-Akyazı, and Irmak Ertör. "The Gezi Park Resistance from an Environmental Justice and Social Metabolism Perspective." *Capitalism Nature Socialism* 26, no. 1 (2015): 99–114.

Öztürk, Murat, Joost Jongerden, and Andy Hilton. "Commodification and the Social Commons: Smallholder Autonomy and Rural–Urban Kinship Communalism in Turkey." *Agrarian South* 3, no. 3 (2014): 337–67.

Paker, Hande. "The 'Politics of Serving' and Neoliberal Developmentalism: The Megaprojects of the AKP as Tools of Hegemony Building." In *Neoliberal Turkey and Its Discontents: Economic Policy and the Environment under Erdogan*, edited by Fikret Adaman, Bengi Akbulut, and Murat Arsel, 103–19. London: I. B. Tauris, 2017.

Pearsall, Hamil, and Isabelle Anguelovski. "Contesting and Resisting Environmental Gentrification: Responses to New Paradoxes and Challenges for Urban Environmental Justice." *Sociological Research Online* 2, no. 3 (2016): 121–27.

Pelivan, Gözde. "Going Beyond the Divides: Coalition Attempts in the Follow-Up Networks to the Gezi Movement in Istanbul." *Territory, Politics, Governance* 8, no.4 (2020): 1–19.

Pellow, David. *What is Critical Environmental Justice*? Cambridge: Polity, 2017.

Pow, Choon-Piew. "Neoliberalism and the Aestheticization of New Middle-Class Landscapes." *Antipode* 41, no. 2 (2009): 371–90.

Quastel, Noah. "Political Ecologies of Gentrification." *Urban Geography* 30, no. 7 (2009): 694–725.

Rancière, Jacques. *The Politics of Aesthetics: The Distribution of the Sensible*. New York: Bloomsbury, 2013.

Ruez, Derek. "'Partitioning the Sensible' at Park 51: Rancière, Islamophobia, and Common Politics." *Antipode* 45, no. 5 (2012): 1128–47.

Sözcü. "AKP'li Belediyeler Borç Dağı Bıraktı." April 7, 2019. https://www.sozcu.com.tr/2019/gundem/akpli-belediyeler-borc-dagi-birakti-4326496/.

Suliman, Adela. "Iceland's Reykjavik Tops Index for Green City Getaways." *Reuters*, April 24, 2018. https://www.reuters.com/article/us-cities-environment-green/icelands-reykjavik-tops-index-for-green-city-getaways-idUSKBN1HV1TO.

Tabak, Seda. "Millet Bahçesi'nin Adı Fiyatları Uçurdu." *Sabah*, August 1, 2018. https://www.sabah.com.tr/emlak/2018/08/01/millet-bahcesinin-adi-fiyatlari-ucurdu.

Toker, Çiğdem. "AKM Millet Bahçesi'nde Neler Oluyor." *Sözcü*, September 13, 2019. https://www.sozcu.com.tr/2019/yazarlar/cigdem-toker/akm-millet-bahcesinde-neler-ouyor-5331236/.

Tucker, Ken. "Politics and Esthetics." *Sociology Compass* 5, no. 8 (2011): 712–20.

Tuğal, Cihan. "Resistance Everywhere: The Gezi Revolt in Global Perspective." *New Perspectives on Turkey* 49 (2013): 157–72.

Turapoğlu, Zehra. "TOKİ 81 İlde 154 Millet Bahçesi Yapacak." *Anadolu Ajansı*, May 22, 2020. https://www.aa.com.tr/tr/turkiye/toki-81-ilde-154-millet-bahcesi-yapacak/1850011.

Turhan, Ethemcan. "'Democracy Happens Where the People Are': Social Conflict, Deliberation, and Youth Perspectives in Post-Gezi Turkey." *Southeastern Europe*. Published online ahead of print, November 30, 2017. https://brill.com/view/journals/seeu/aop/article-10.1163-18763332-000010/article-10.1163-18763332-000010.xml.

Turhan, Ethemcan, Semra Cerit Mazlum, Ümit Şahin, Alevgül Şorman, Arif Cem Gündoğan. "Beyond Special Circumstances: Climate Change Policy in Turkey 1992–2015" *WIREs Climate Change* 7, no. 3 (2016): 448–60.

TURKSTAT. "Yapı İzin İstatistikleri, Ocak-Eylül, 2020." *TURKSTAT Haber Bülteni*, no. 33782. November 17, 2020. https://data.tuik.gov.tr/Bulten/Index?p=Yapi-Izin-Istatistikleri-Ocak-Eylul,-2020-33782.

Weise, Zia. "Istanbul's Bid to Become Green Capital of Europe 'Is a Joke'" *Guardian*, November 3, 2014. https://www.theguardian.com/environment/2014/nov/03/istanbuls-bid-to-become-green-capital-of-europe-is-a-joke.

Wolch, Jennifer R., Jason Byrne, and Joshua P. Newell. "Urban Green Space, Public Health, and Environmental Justice: The Challenge of Making Cities 'Just Green Enough.'" *Landscape and Urban Planning* 125 (2014): 234–44.

World Bank. "GDP Growth (Annual) Turkey." *World Bank Open Data* https://data.worldbank.org/indicator/NY.GDP.MKTP.KD.ZG?end=2020&locations=TR&start=1998 (accessed 31 May 2021).

Yavuz, Yusuf. "İklim Zirvesinde Millet Bahçesi Utancı." *Oda TV*, September 24, 2019. https://odatv4.com/iklim-zirvesinde-millet-bahcesi-utanci-24091925.html.

Yeni Şafak. "Erdoğan'ın Açıklamasıyla 8 Mahallede fiyatlar uçtu." July 31, 2018. https://www.yenisafak.com/ekonomi/erdoganin-aciklamasiyla-8-mahallede-fiyatlar-uctu-3387117.

Yetişkul, Emine, and Şule Demirel. "Assembling Gentrification in Istanbul: The Cihangir Neighbourhood of Beyoğlu." *Urban Studies* 55, no. 15 (2018): 3336–52.

Yıldız, Emrah. "Cruising Politics: Sexuality, Solidarity, and Modularity after Gezi." In *The Making of a Protest Movement in Turkey*, edited by Umut Özkırımlı, 103–20. London: Palgrave, 2014.

Yörük, Erdem. "The Long Summer of Turkey: The Gezi Uprising and Its Historical Roots." *South Atlantic Quarterly* 113, no. 2 (2014): 419–26.

Zengin, Aslı. "What Is Queer about Gezi?" Society of Cultural Anthropology, October 31, 2013. https://culanth.org/fieldsights/what-is-queer-about-gezi.

Breaking Consensus, Transforming Metabolisms

Notes on Direct Action against Fossil Fuels through Urban Political Ecology

Salvatore Paolo De Rosa

The impacts of climate breakdown are increasing their toll, hitting hardest the countries and communities least responsible for climate-changing emissions.[1] The facts have never been clearer,[2] but facts do not speak for themselves. Scientific consensus does not, and cannot, establish a clear-cut, universal political consensus. To acknowledge that the climate is changing due to anthropogenic activity is different from determining which experiences and understandings of climate change—in its causes, effects, and solutions—ultimately matter.[3] This is the realm of politics: a field traversed by a multiplicity of standpoints and structured by unequal relations of symbolic and material power. Therefore, a consensus-based conception of the political and a singular conception of climate change may prevent the articulation of conflicting interests and of competing visions that do not fit dominant framings and may end up in reproducing hegemonic narratives thriving on the "invisibilization" of disagreement.[4]

Debates around climate mitigation exemplify such tensions. Scientists agree that deep decarbonization is unavoidable to mitigate global warming, but while "cost effective" gradual mitigation pathways and faith in technological solutions are part of a global consensus around climate change governance,[5] decisive reductions in fossil fuels use—the main culprit of greenhouse gas (GHG) emissions—through supply-side policies have not gained traction and are still considered politically challenging.[6] In this context, a rising wave of climate justice activism is mobilizing to decommission emitting devices, establish a moratorium on new fossil energy projects, and leave

Social Text 150 · Vol. 40, No. 1 · March 2022
DOI 10.1215/01642472-9495160

fossil fuels in the ground.[7] Just underneath the thin cover of consensus lies the conflict attempting to break through.

According to the geographer Erik Swyngedouw,[8] the consensual regime of climate change governance is an outcome of the current post-political condition that attempts to foreclose politicization and evacuate dissent through apparent participation and technocratic expertise in the context of an undisputed market-based socioeconomic organization. Under this condition, the work of climate justice movements becomes more relevant than ever, for they attempt to disrupt the regime of invisibility in which alternative values, desires, experiences of, and solutions to climate change are relegated, by bringing back dissent as the true engine of politics.[9]

Climate justice movements draw attention to the historical and contemporary inequalities that underlie climate change and make those least responsible for emissions the most affected by climate chaos, advocating for mitigation and adaptation strategies that do not increase vulnerabilities but reduce them, therefore rejecting false solutions.[10] Since the late 2000s, these movements have engendered a new political space for the critique of official climate policies and have engaged in various forms of direct action, providing a platform for antisystemic approaches against the elite capture of the climate debate.[11] At its core, climate justice involves an antagonistic framing of climate politics that breaks with attempts to construct climate change as a "post-political" issue[12] and instead targets the political process itself with an agenda of systemic transformations. Thus, their approach holds the potential to embrace the urgency of ending the fossil fuel era while retaining commitments to justice and equality. But which kind of strategies and tactics are best positioned to counteract the traps of consensus and to forward just transformations?

In this essay, I begin exploring this question through an analysis of what I consider one of the most promising current instances of the climate justice movement: direct action against fossil fuels' infrastructure, manufacture, and development. Drawing on the conceptual toolkit of urban political ecology (UPE), I frame the debate around solutions to global warming away from a consensus-based conception of the political to bring to center stage multiple aspirations for metabolic transformations of the socioecological assemblages and imaginaries produced by, and sustaining, the capitalist economy. On this basis, I reflect on the case of the Swedish climate justice movement Fossilgasfällan (the Fossil Gas Trap, FGF) through UPE, focusing in particular on the blockade of the Gothenburg gas terminal in 2019. My aim is to provide some initial notes on the potential of direct action against fossil fuels to break consensus around useless and unjust solutions and to enact an effective and transformative climate justice politics.

An Emerging "Blockadia"

In the thematic map "Blockadia," the Environmental Justice Atlas reports sixty-nine ongoing cases of resistance movements and place-based mobilizations against fossil fuel projects along the whole chain, from extraction to transportation to combustion.[13] The term Blockadia, popularized by Naomi Klein,[14] refers to this transnational, loosely connected conflict zone where the defense of land, livelihoods, and climate is expressed through direct actions such as blockades, occupations, and protests. Local resistance against fossil fuel extraction is not new. In a review based on the atlas, Temper and coauthors record 371 cases of conflicts related to fossil fuels from 1997 to 2019.[15] A notable antecedent to the current surge is the fight of the Ogoni people of the Niger Delta against Shell's operation in the 1990s,[16] culminating in the withdrawal of Shell from Ogoni land. What is new is the growing awareness of the interconnectedness of these spaces as multiple frontlines in a global struggle for socioenvironmental and climate justice.

Besides defending local livelihoods, direct action against fossil fuels can be seen as a supply-side strategy of climate mitigation from the bottom up.[17] According to Piggot, "analysis suggests that a large portion of global fossil fuel reserves will need to remain unburned to keep climate change 'well below' 2°C. . . . Yet, investment in fossil fuel infrastructure continues at a pace that is inconsistent with agreed climate goals, and no meaningful global policies exist to keep fossil fuels in the ground."[18] This has been the result of an organized pushback by the fossil fuel industry, employing a variety of tactics to keep its business afloat, including political influence, climate change skepticism, and co-option of local groups.[19] While neither the European Green Deal (EGD) nor the European Commission energy plans promote the expansion of fossil fuels for climate reasons, the latest European Union list of prioritized energy infrastructure projects—the Projects of Common Interest (PCI), benefiting from simplified permissions and EU funding—includes thirty-two fossil gas projects eligible for funding of up to twenty-nine billion euros. Since the PCI lists were introduced in 2014, 42 percent of the total amount of funding has gone to fossil gas infrastructure projects.[20] Moreover, as a recent report shows,[21] governments are planning to produce about 50 percent more fossil fuels by 2030 than would be consistent with limiting warming to 2°C and 120 percent more than would be consistent with limiting warming to 1.5°C.[22]

Resistance is thus mounting. In recent years, in North America, grassroots coalitions of Indigenous Nations and environmentalists animated the struggles against the Keystone XL, Dakota Access, and Trans Mountain pipelines, catalyzing mass movements of international resonance that have been brutally repressed by state forces.[23] In Europe, a varied and interconnected landscape of direct-action coalitions against fossil fuels has emerged during the last decade, building on the myriad local conflicts against fossil fuel extraction and infrastructure. Some of the most prominent coalitions include Ende Gelände (Here and No Further), one of the biggest direct-action climate coalition groups blocking coal mines and coal-fired power plants in Germany since 2015; Coode Rod (Red Code), a climate justice group from the Netherlands that amplifies the resistance to gas extraction in Groningen and is the main instigator of the global campaign Shell Must Fall, and Frack Off! in the United Kingdom, which was instrumental in the government's suspension of gas extraction on the British Islands in 2019. Despite growing membership and some victories, these movements are also increasingly criminalized and framed as extremists,[24] becoming the target of state repression through violence,[25] espionage,[26] and counterinsurgency techniques.[27]

What unites such diverse coalitions is the reaction against the occupation of space (both earthly and atmospheric) and of the future provoked by the relentless extraction and burning of fossil fuels. Their repertoire of action includes campaigns for divestment[28] and for legislation against fossil fuels.[29] Alongside these actions, or following their failure, a crucial tactic remains the blockade: a counter-occupation that, besides denouncing the inconsistencies and the hypocrisy of governments and corporations, aims to repoliticize fully the debate around climate change by providing platforms for radical critiques of the status quo. A blockade is an occupation for liberation through which coordinated movements counteract false solutions and the evacuation of justice concerns, call for wide-ranging transformations, and embody visions of alternative futures in the practice of the camp.[30] In the following section, I discuss key concepts of UPE in relation to the approaches and politics of direct action against fossil fuels. My argument is that these activists display a metabolic perspective of capitalist-driven environmental destruction and global warming, prompting a radical critique of the consensual regime of climate politics through political performances aimed at metabolic transformations.

Metabolic Activism: UPE and Direct Action against Fossil Fuels

The manifold socioecological processes at the root of global warming, and the uneven distribution of their effects among different geographies and social groups, are inextricably linked to the urbanization of nature. Drawing upon UPE, I define the urbanization of nature as the relentless transformation and mobilization of biophysical entities for feeding the expansion of the urban form on a planetary scale, a process unfolding through the social and material relations organized by the dominant economic system of neoliberal capitalism.[31]

UPE scholars investigate urbanization through the prism of socioecological metabolisms: the dynamic assemblages of political economic processes and biophysical transformations along global networks that make up the concrete realities of cities.[32] A metabolic understanding of nature's urbanization brings to the forefront the ways in which societies incorporate biophysical entities into their functioning, providing insights into the role of power distribution at several scales in determining the drivers, the forms, and the outcomes of socioenvironmental change. Through this perspective, the analysis of commodities, such as fossil fuels, reveals the processes that turn a material thing into a resource, uncovering the connections between different histories and geographies, the main actors and nodes, the socioenvironmental effects and the unequal power relations embedded in metabolic assemblages.

The process of urbanization has never been solely about the shape of the city; neither is it contained within city borders. On the contrary, urbanization has always been about what happens *beyond* urban agglomerations, as the myriad processes fundamental to the building and functioning of cities shape the operational landscapes of global production and supply chains stretching across continents.[33] These *extended* forms of urbanization—sites of resource extraction, agro-industrial production, energy and information circulation, waste management, and military occupation—wrap the world in interwoven networks of global metabolic exchanges.[34] The places in which such sites, routes, and nodes integral to cities have proliferated are frequently the setting of environmental conflicts waged by impacted communities against processes detrimental to their survival. Alongside local communities, direct-action movements for climate justice are increasingly targeting the ramifications of fossil energy hidden in plain view: the wells and drills through which fossil fuels are extracted; the pipelines, roads, railways, airports, and ports that allow their circulation; and the petrochemical factories that multiply their pervasiveness. The primary aim is to stop them from continuing their activities through the force of social cooperation. Civil disobedience provides the broader framework and the connection with a long genealogy of grassroots action to stir change.[35]

Another crucial insight of UPE is an understanding of the political process as fundamentally shaped by conflict. UPE's approach focuses on the question of who gains from and who pays for particular trajectories of socioenvironmental change articulating at several scales.[36] By resorting to physical and discursive strategies of resistance and self-organization of alternatives, impacted communities and activists disrupt the active depoliticization of social antagonisms and socioecological metabolisms, counteracting both the material processes producing socioenvironmental inequalities and the hegemonic narratives of environmental problems and solutions. This is particularly relevant in light of current climate change governance as a manifestation of the postpolitical condition that forecloses politicization and evacuates dissent by rejecting ideological divisions and by reducing the political terrain to the sphere of consensual governing and policymaking within the given neoliberal order.[37] In practice, this has allowed the capture by elites of the framing and management of the climate crisis. Conversely, according to UPE scholars, the remaking of socioecological relations under global warming should be framed not as a techno-managerial problem but rather as a metabolic transformation of the socioenvironmental assemblages underpinning the urbanization of nature through political performances that make visible multiple potential futures beyond capitalism.

Conflict and dissent, in this perspective, are understood as genuine manifestations of "the political": a performative interruption of the policed order of the sensible by the part that has no part.[38] This conception is in line with agonistic theories of climate change,[39] which consider consensus undesirable because it suppresses dissent and embodies a rejection of the political. Consensus, rather than merely suggesting agreement, is also an expression of hegemony, which subordinates some political identities to others.[40] Therefore, it is only through multiple forms of counter-hegemony that a radical democratic praxis of climate change may flourish.

UPE's approach resonates with climate justice activism and in particular with grassroots battles waged against fossil fuels. My contention is that they share a metabolic perspective insofar as they both assign preeminent political relevance to the transformation of the flows, nodes, and spaces structuring the capitalist urbanization of nature—of which fossil fuels constitute the *lifeblood*—not only to halt global warming but also to rebuild a more just and equal world. Related to that, both aim to debunk and discard the imaginaries of society-nature relations that sustain these assemblages in order to open up the array of interpretations of, and solutions to, climate change to marginalized voices and experiences. Therefore, I call for increased conceptual scrutiny and political support toward direct action against fossil fuels, a form of climate justice activism currently at the fringes of the climate movement but that reconnects with a long history of physical and symbolic occupations aimed at radical change.[41] In so doing, I answer to recent calls by critical urban scholars toward "developing a vocabulary and grammar that can support politically performative theory and engaging with real existing political movements."[42] Preliminarily, I define as *metabolic activism* those instances of grassroots ecopolitical engagement that aim to disrupt, block, occupy, and ultimately transform capitalist-driven metabolic flows and relations by intervening directly in the operational landscapes that quilt together these assemblages while experimenting with alternative values, knowledges, spaces, and sociomaterial relations.

Grounded in a conflictual attitude toward the status quo, metabolic activists go beyond the politics of invitations to powerful elites by enacting confrontational tactics in alliance with civil society, impacted communities, and grassroots environmental movements. They deploy blockades and occupations to confront "carbon entanglements"—the deep interconnection of economies and political structures with the fossil fuel industry[43]—and rise against the networks and actors driving the world toward climate chaos while pretending to roll out solutions.[44] Their activism is a reaction to a depoliticized public sphere and a tool of repoliticization, staging those demands, experiences, and desires banned from the consensual order of climate change governance. It is precisely this order that allows fossil fuel companies and investors to present their techno-managerial fixes as green, sustainable, and climate-friendly initiatives, while relentlessly continuing extraction, multiplying emissions, and locking in increasing global warming. In the attempt to foster radical transformations, direct-action climate coalitions lay siege to the consensus around technological, market-led, and growth-focused transition by doing the "dirty work" of stopping the machine with their bodies. Not by focusing on corporations' headquarters and government's buildings in city centers but by going "out there," in the sparse geographies of fossil fuels' extraction, flows, and manufacture. Moreover, and crucially, metabolic activists for climate justice seek to recast the climate change debate from just an issue of emissions' accounting to a general critique of the hegemonic socioeconomic, political, and cultural systems built on environmental destruction, class divisions, colonial legacies, racism, and imperial relations. To ground and expand these reflections, in the following section I trace the rise and the struggle of Fossilgasfällan (FGF), the Swedish climate justice movement that organized the first-ever blockade of fossil fuel infrastructure through direct action in Sweden.

Fossilgasfällan: Direct Action against Fossil Gas

In recent years, the port of Gothenburg in Sweden has stepped up ambitions to become a hub for storage and selling of so-called liquefied natural gas (LNG). On May 30, 2014, the administrative board of Västra Götaland County granted to Swedegas, the infrastructure company that owns and runs the Swedish gas transmission network, a permit for the expansion of the LNG terminal in the port of Gothenburg under the project name Go4LNG. The permit allowed for increasing storage capacity up to thirty-three thousand cubic meters of LNG and for handling a maximum of five hundred thousand tons of LNG per year. A crucial leg of the project aimed to link the terminal to the national gas network directly, fostering increased and long-term gas imports to supply the country's energy grid. However, this connection did not materialize. On October 10, 2019, the Swedish government rejected the Swedegas application because of climatic concerns: according to the Minister of the Environment and Climate Isabella Lövin, the project of connecting the terminal with the national grid contradicted Sweden's ambitious national climate goals, as it would have locked the country into fossil infrastructure.[45] In between the approval and the rejection, fundamentally influencing the debate in the country and the final decision, were three years of grassroots campaigning spearheaded by Fossilgasfällan, a Swedish climate justice movement that arose in opposition to the consensus surrounding gas as a clean and climate-friendly energy source.

A small group of Swedish activists based in Gothenburg founded FGF in 2017. The founding members had a background in environmental conflicts, animal rights, migrant justice, Sámi rights, and environmental NGOs. Some of them were involved for years in *Fossil Free Sweden*, a campaign advocating divestment from fossil fuels by public and private institutions, without much success in Gothenburg. The choice of targeting fossil gas also reflected their relation with the Gastivists, a global network devoted to fight the normalization of fossil gas as a "transition fuel" by connecting frontline communities with climate justice groups to oppose new gas infrastructure. The Gastivists network facilitates knowledge creation and sharing, and provides a practice-based model of intervention rooted in civil disobedience through workshops and trainings.

The initial work of FGF focused on building a knowledge base of the relations between climate change and fossil gas. Its main point of contention was the framing in policy circles and in the energy sector of gas as a bridge fuel for the energy transition required by climate change. Starting from the name—dismissing "natural" for "fossil" gas—its analysis deconstructed the rising popularity of gas, showing how this was the result of a clever greenwashing operation by fossil fuel companies and their lobbies in government chambers.[46] The main tools for building the campaign were social media, newspaper articles, and actions in cities. Through a dedicated website, FGF set up a "knowledge bank" collecting sources challenging fossil gas and instructing on alternatives. The social media channels of FGF on Facebook and Twitter were instrumental for reaching a large audience, alongside physical venues to spread flyers and posters with the help of sympathetic campaigns and movements. Virtual info meetings on YouTube provided an easy way to broadcast information, to gather support, and to expose the financial interests and players behind the Swedegas project. Informed by climate justice concerns, FGF took seriously the claims and experiences of local communities impacted by the extractive processes to harness fossil gas, most notably through hydraulic fracturing, also known as "fracking." This led to a holistic critique of fossil gas associated with the development of a set of alternative proposals for the transformation of energy systems and of society as a whole.

In one of FGF's first public actions, on February 11, 2018, about thirty people gathered outside the Social Democrats' office in Gothenburg to protest the construction of the terminal Go4LNG that had begun despite the fact that Swedgas lacked a concession under the Natural Gas Act for the entirety of the project. In a fiery speech, activists, besides denouncing the false promises of gas, set the terms of the issue within the broader geographies and consequences of extraction:

> Fossil gas has major negative impacts where the gas is extracted and infrastructure is built: water poisoning, earthquakes, landslides, repression of human rights, destruction of land areas and agricultural areas. What is happening now in Gothenburg harbor affects all of Sweden and all of the world's people.[47]

The campaign by FGF gave the climate movement in Sweden a clear target around which a broad alliance of activists and concerned people could coalesce. To ground complex and seemingly abstract issues like climate change by targeting specific nodes and facilities implicated in the fossil economy, where opportunities for short-term gains and victories sustain momentum and movement building, has been an explicit strategy of the climate movement for at least a decade.[48] This approach has also helped reconnect with the historical legacy of direct-action environmental campaigns and with contemporary struggles of frontline communities against fossil fuels extraction and infrastructure. In particular, looking at the past, the antinuclear German movement blocking trains carrying nuclear fuel and waste during the 1980s represents the clearest European antecedent to FGF, and can be seen as the precursor inspiring Ende Gelände.

Moreover, by linking ideally and practically with frontline communities resisting fossil fuel operations and expansion, the movement against fossil gas in Sweden was able to frame its actions as both an ecological and social battle with immediate implications in the life of communities far away from Sweden, and as a decisive step to halt GHG emissions. This internationalist approach was operationalized by inviting to Sweden representatives of communities impacted by gas extraction and infrastructure from the United States, Argentina, and Ireland. A counter-narrative of gas metabolism was elaborated by merging the claims of place-based campaigns for environmental justice, Indigenous sovereignty, and rights to a clean environment with the climate struggle against the expansion of fossil fuels in Sweden. A series of video interviews became the medium to advance this narrative, highlighting the injustices along the whole gas metabolic chain, and linking campaigns through solidarity and mutual support.[49]

The first aim of FGF was to put the task of debunking fossil gas as climate-friendly transition fuel on the agenda of the climate movement and of Swedish climatic debates. From this aim descended the work of self-formation and knowledge sharing, the linking with antiextraction campaigns around the world, and the subsequent targeting of public media discourse. Its second aim was to block the completion of the terminal, considered unjustifiable for three interrelated reasons. First, LNG may reduce carbon dioxide emissions, but it tremendously increases methane emissions, a powerful GHG;[50] second, gas infrastructures linked to the national energy grid lock in fossil energy imports (in Gothenburg it would have been for about forty years); third, a fossil "transition" fuel is not a step toward comprehensive social and ecological transformation but rather an obstacle to it. Finally, through this campaign, activists pursued the strengthening of the climate justice movement in Sweden and Europe by learning from each other and by building on and making use of the momentum they collectively created (from an interview with A.F., activist in FGF).

The strategic approach of FGF relied on what it called a "staircase model," meaning a gradual deployment of tactics that escalate each time the movement's claims and requests go unheard. This unfolded with various forms of dialogue to drive media and political attention, then more confrontational actions, as the movement grew and the project continued (like a spectacular, if rather windy, kayak demonstration in 2018) and culminated in the coordinated mass direct action to block the terminal. The latter became inevitable when the project continued its course toward the final permit. At the beginning of 2019, according to FGF, an intervention to influence the pending decision by the government was needed. It had already experimented with direct action in December 2018, when a group of thirty people demonstrated for a few hours in front of the gate of the terminal. The time had come to raise the bar. Together with the growing network of European direct-action movements against fossil fuels, supported by the international coalition Gastivists, and enlisting social justice movements from all of Sweden, FGF participated in the creation of Folk mot Fossilgas (People against Fossil Gas), which was simultaneously a coalition, an event, and a climate camp geared at blocking the gas terminal's operations for at least one day during September 6–8, 2019.

Folk mot Fossilgas, the Camp, and the Blockade

About six months before the designated days, the campaign Folk mot Fossilgas was launched through an international call inviting "anybody who wants to act for climate justice and who agrees with the action consensus" to join the Swedish activists at the port of Gothenburg. The action consensus is a typical mode of involvement utilized by direct-action movements against fossil fuels in Europe to ensure an informed participation by anybody who is sympathetic with the movement's goals and tactics, and to allow the widest participation irrespective of strict ideological adherence to a specific movement's identity.[51] The kind of consensus sought by climate and environmental justice activists does not silence antagonism and partisanship. Instead, it exposes them by drawing a line against racial, ethnic, and gender discriminations and by grounding collective action in antifascist and anticapitalist values.

The campaign was organized around eight working groups—action planning, finance, mobilization, communication, logistics, training, well-being, and legal issues—each with two facilitators serving as coordinators. The organizers chose to have a relatively high degree of anonymity and security during preparations, using code names, encrypted communication, and technology-free in-person meetings. This implied a careful navigation of the complex balance between agility and safety. As the date approached, the European network of allies cooperated in the organization of meetings and trainings in Copenhagen, Berlin, Amsterdam, Brussels, and other northern European cities. These encounters recruited participants while the trainings introduced the neophytes to civil disobedience and to the organizing principles of coordinated direct action. The trainings represent a crucial element for the success of the action, transferring practical skills and building a sense of empowerment and safety among participants, preparing them to respond to police intervention through nonviolent civil disobedience and providing coordinating skills to manage decisions within affinity groups that try to dismantle groups' internal power dynamics.

On the morning of September 6, 2019, people began converging on the camp set up by the organizers in a meadow about one kilometer from the Gothenburg port. The camp comprised a big circus tent for the assemblies plus five smaller structures housing the info point, the area to eat meals, the media center, the legal support, and the health support. Next to the seashore, about a hundred camping tents were distributed into rows. The media center worked tirelessly to film, edit, and upload content, to write up articles and press releases, and to scan the mainstream media narrative of the event before, during, and in the aftermath of the blockade. Interviews with international delegates in video and in crowded plenaries brought to Gothenburg the concerns and the solidarity of frontline communities fighting against gas extraction and gas terminals. One of them was "Christopher" an Esto'k Gna from the tribe of Carrizo Comecrudo that dwells in the valley of the Rio Grande in Texas. Alongside the tribe's reservation, three different shale gas extraction sites had been approved, causing earthquakes and polluting groundwater. That gas was then shipped to European terminals such as the one in Gothenburg. According to Christopher, fighting the extraction and trade of fossil fuels is just the continuation of the anticolonial struggle of Indigenous peoples against a process of exploitation and genocide that began in 1492.[52]

The most important purpose of the communication team and the media center was to make sure that the narrative of the campaign reached out to the broader public. Since mainstream media and politicians often dismiss civil disobedience, the "narrative battle" was a key site of resistance. Activists led it highly effectively, as shown by the fact that spokespersons from the campaign were interviewed in most Swedish major media and that the action was mentioned by the media in relation to the government decisions a month later to stop the terminal. Special care was also devoted to the well-being of participants, through a dedicated group of people with and without medical experience who provided physical and mental support to keep the activists in good shape. They offered workshops on sustainable activism during the camp, help with first aid kits, blankets, water, and snacks (and also hugs and kind words) during the action, and conversational support to activists who needed it after the action.

The practical tasks to keep the communal life of the camp going were distributed on a voluntary basis and taken very seriously by volunteers. The average age was under thirty and the overwhelming majority European and white. Time passed between assemblies, food sharing, and trainings. The trainings were organized by the group Climate Justice Program and were meant to introduce neophytes to the tactics of civil disobedience and coordinated mass action, the distinctive tools of the militant climate movements consolidating in Europe. Essentially, these kinds of actions are mass interventions on a physical infrastructure implicated in the fossil economy with the aim of slowing it down or blocking it altogether. The objective is not a sabotage but a demonstration of grassroots force that aims at maximum visibility to nurture imaginaries of revolt and to inspire emulation. As acts that defy the law—invading private property or occupying public places—the risk of being arrested is an integral part. The high number of participants guarantees collective security. Violence against things and provocations are banned, but other forms of illegality connected to the success of the action, and carried out in a group, are considered legitimate. The training sessions are necessary to socialize the complex series of organizational principles and communication codes that allow dealing with highly dynamic actions in a coordinated manner, responding quickly to changing conditions.

The attention to forms of dialogue that enhance horizontal decision-making, before and during the action, not only responds to the refusal of authoritarianism but also to the attempt at "erasing" the leaders from the picture, thus avoiding that certain subjects more exposed become visible to repression. The micro-techniques that safeguard collective consensual action owe to the political awareness of militant feminism. Starting from the concentric structure—which has buddies at its core, integrated in affinity groups, then in action fingers with different roles, in turn collected in the totality of the mass—up to the sign language and the facilitation techniques that lead the discussions, without neglecting spaces dedicated to the expression of individual emotions and self-criticism, the techniques are, overall, a set of practices rooted in the tradition of civil disobedience but conjugated in new forms. When it goes well, the result is the collective construction of a symbolic event, highly visible and easily understandable to the outside, inviting and safe for inexperienced activists, culminating in a formative and transformative practice for individual participants.

The exact plan of how the blockade would unfold remained secret to avoid preventive police boycott and was arranged by a core group of activists who studied the terrain and the targeted terminal for months. The next morning, about five hundred people were divided into three fingers ready to block the terminal. Halfway to the terminal, the fingers separated toward three different gates. The finger I joined arrived in front of the electrified gate of the main entrance. The space was conquered easily by sitting close together on the pavement. A group of about twenty police officers stood still. On both sides of the gate, the row of tank trucks was blocked. In the meantime, drones were buzzing around for filming by the action organizers while at least two other people were taking photos until the evening. These were professional photographers and filmmakers, chosen to ensure good visual communication around the campaign. Their images later became the media product disseminated online, part of the strategy to influence the narrative on legitimate practices of climate resistance and to present this kind of action as exciting and effective.

Two people in yellow vests had the role of communicating with the police. After a brief exchange, they spread the news that the police were not going to remove us. The tension disappeared as the same news arrived from the other blocks. Everything ran smoothly, the terminal was blocked, the trucks did not enter or leave, the police merely observed, and the morale was high. Several television stations carried out interviews, showing the blockade on major Swedish media and talking about "fossil gas" instead of "natural gas," a clear symbolic victory. At noon, the food brigade arrived to feed us, going to all the gates and coming back for dinner. In the end, the blockade lasted more than twelve hours, achieving the interruption of the terminal's output flow, positive media coverage, and no violence from the police, this time.

While it would be a mistake to overemphasize this single action as the cause of the government rejection of the final permit for Swedegas, it certainly managed, together with three years of campaigning, to occupy and to fracture in multiple ways the widely held conventional wisdom of Sweden as climate leader. Further, it represented a turning point for the climate justice movement in this and neighboring countries: coalition-building, direct action, and blockade were shown to be viable strategies to break consensus and to influence the course of national climate politics.

An Emerging Metabolic Activism?

Looking at direct-action movements against fossil fuels like FGF through UPE's conceptual toolkit can help illuminate their relevance in the current conjuncture. Here, building on previous sections, I chart three interrelated points that provide a basis for further investigations.

A first element is that such movements disrupt the hegemonic consensus around top-down climate governance by showing its inconsistencies in order to make space for other claims, knowledges, and experiences. In this sense, they react to a postpolitical condition in which climate change has been taken up as an object of regulation and policy, but only by evacuating the structural and systemic inequalities that are at its root and that are reflected in the uneven socioecological outcomes of climate breakdown and the "green transition."[53] They do so by critically addressing both the specific set of discourses and interventions put in place by states, corporations, and international agreements, and the political and economic arrangements within which these actors build their legitimacy and ground their power. By bringing back dissent as the ultimate engine of truly democratic politics, these movements repoliticize the debate around climate solutions and link it directly to the struggles for land, water, and other basic socioenvironmental needs waged by impacted communities. FGF enacted such politics by contesting the legitimacy of gas as a transition fuel endorsed by the Swedish state, putting the spotlight on the entanglement of financial interests and socioenvironmental consequences of fossil gas along all stages of its metabolism.

This is linked to the second point: the metabolic perspective detectable in direct-action movements against fossil fuels allows them to shift the focus of climate politics onto the financial players and material infrastructures that reproduce the relation between fossil economy and climate chaos. From the hegemonic consensual framing of climate action as the management of disembodied emissions—a leap that has helped the abstraction of real emissions into rates and amounts that could be traded and/or offset[54]—the targets become the concrete actors and material nodes of the fossil fuel economy, thus welding emissions to a larger set of concerns, including neocolonial relations, habitats, and local environments destruction, class and gender issues, pollution and health, lack of democracy, and the prospect of transformations toward a fossil fuel–free society. In this sense, FGF's aim has been to expand the locally focused Swedish debate around gas in light of the social and environmental articulations of gas as energy source and GHG, globally and historically. The metabolic perspective is visible also in FGF's strategy: based on a deep engagement with the local context and on thorough knowledge of Swedish climate politics, and at the same time harnessing resources, support, and legitimacy from international networks of climate justice and grassroots movements. In its approach, pursuing the goal of stopping one specific fossil gas project is inseparable from the global struggle against all fossil fuels. By making the tensions between the local and the global productive of relations, claims, and strategies, FGF is best positioned to tackle an issue that is simultaneously global and local such as climate change.

Third, by (re)creating spaces and communities around resistance to the nodes of the fossil fuel metabolism of capitalist economies, direct actions turn climate justice concerns into concrete practices of transformation and prefiguration. Indeed, the present iterations and uses of blockades and occupations constitute a formidable laboratory of grassroots climate politics that links the struggles of frontline communities with the organizing effort of urban activists. The critique of fossil-powered capitalism makes clear that solutions will not come from the elites, prompting the search for alternatives that would be democratically managed and socioecologically viable in the broadest sense. Indeed, FGF explicitly proposed a vision of fossil-fuel-free society to articulate through the banning of all fossil extraction, infrastructure, and imports; the massive reduction in energy consumption; the creation of a wider welfare for all; and by making citizens energy producers rather than just energy consumers. Moreover, the everyday life of the camp puts into question the pursuit of self-interest and competition (hallmarks of neoliberalism) in favor or practices of mutualism and cooperation. The experience of participation can be potentially transformative for individuals insofar as it renders tangible in practice alternative social and ecological relations, making room for forms of interaction and values that disrupt dominant epistemologies and subjectivation processes.

Conclusions

According to UPE, the current postpolitical condition reduces the space of politics to techno-managerial governing and policymaking within the given neoliberal order, thus foreclosing a radical politicization of the status quo. This is particularly evident in the current hegemony of a consensual regime of climate change governance revolving around market mechanisms and technological upgrades that do not fundamentally challenge either the drivers or the systemic inequalities at the root of environmental destruction and global warming. In this context, dissent and antagonism become fundamental to break up the consensus and to make visible marginalized interpretations of, and solutions to, climate change. In this article, I argued that direct-action climate justice movements against fossil fuels may be the best positioned to counteract the traps of consensus and to forward just transformations, while retaining commitments to fairness and equality. I provided some initial notes on the relevance of their strategies and tactics through the framework of UPE, grounding my analysis in the empirical case of FGF, a Swedish climate justice coalition fighting against the development of fossil gas. Direct-action movements against fossil fuels share with UPE a metabolic understanding of global socioecological systems that assigns political relevance to the transformation of the flows, nodes, and spaces structuring the capitalist urbanization of nature in order to halt global warming and to rebuild a more just and equal world. Moreover, they both abide by a conception of the political based on conflict and dissent, promoting multiple forms of counter-hegemony through a radical democratic praxis of climate change.

Direct-action climate justice movements against fossil fuels deploy counter-narratives, blockades, and the camp as primary tactics of their political work, reconnecting with a long tradition of civil disobedience. In Sweden, FGF, through the Folk mot Fossilgas coalition and event, has been able to mobilize five hundred people in a direct action against the expansion of the gas terminal in Gothenburg. As a culmination of FGF's three-year-long campaign directed at Swedish society, the blockade has contributed to the dismissal of the terminal expansion by the Swedish government, crucially influencing the public representation of fossil gas and materially halting the terminal operations for one day. Such actions and related campaigns are fundamental to break the hegemonic consensus around top-down climate governance. Besides showing its inconsistencies and its capture by the elites, they make space for other claims, knowledges, and experiences, shifting the focus of climate politics to the financial players and material infrastructures that reproduce the relations between fossil economy and climate chaos, and turning climate justice concerns into concrete practices of transformation and prefiguration.

These initial considerations warrant more sustained and engaged inquiries into direct-action activism against fossil fuels through metabolic and political-ecological perspectives. Important issues remain understudied, such as the relations between direct action and violence, the potential for escalating and generalizing this activist strategy, and the prospect of alliances between movements against fossil fuels and movements of workers in the fossil fuel sector. It is timely for scholarly analysis to seek a better understanding of venues for transformation and of the best strategies and tactics to build a more just and equal world. In this sense, looking at direct action against fossil fuels may provide a promising and politically relevant field of inquiry.

Salvatore Paolo De Rosa is a postdoctoral researcher at Lund University Centre for Sustainability Studies, Sweden, and an activist for climate justice. His background is in anthropology and geography, while his current research focuses on the political ecology of climate movements.

Notes

Support for work on this article was provided by FORMAS (Swedish Research Council for Sustainable Development) under the National Research Programme on Climate (contract 2017-01962_3) and under project 2018-02800, "Global Attribution Models, Mediation and Mobilisation (GAMES)."

1. Bathiany et al., "Climate Models."

2. Höök and Tang, "Depletion of Fossil Fuels"; Sippel et al., "Climate Change Now Detectable."

3. Machin, *Negotiating Climate Change.*

4. Kakenmaster, "Articulating Resistance."

5. Carton, "Carbon Unicorns"; Watt, "The Fantasy of Carbon Offsetting."

6. Lazarus, Erickson, and Tempest, "Supply-Side Climate Policy."

7. Malm, *How to Blow Up a Pipeline*; Temper et al., "Movements Shaping Climate Futures"; Piggot, "The Influence of Social Movements."

8. Swyngedouw, "Apocalypse Forever?"

9. Kenis and Lievens, "Searching for 'the Political.'"

10. Walker, *Environmental Justice.*

11. Rosewarne, Goodman, and Pearse, *Climate Action Upsurge.*

12. Chatterton, Featherstone, and Routledge, "Articulating Climate Justice."

13. Environmental Justice Atlas, "Blockadia."

14. Klein, *This Changes Everything.*

15. Temper et al., "Movements Shaping Climate Futures." The review also reports 278 conflicts related to low-carbon-energy projects.

16. Bassey, "Leaving the Oil in the Soil."

17. Le Billon and Kristoffersen, "Just Cuts for Fossil Fuels?"

18. Piggot, "The Influence of Social Movements," 1.

19. Misutka et al., "Processes for Retrenching Logics."

20. Nace et al., "The New Gas Boom."

21. SEI, IISD, ODI, Climate Analytics, CICERO, and UNEP, *The Production Gap.*

22. Tong et al., "Committed Emissions from Existing Energy Infrastructure."

23. Estes, *Our History Is the Future*; Cheon and Urpelainen, *Activism and the Fossil Fuel Industry.*

24. Nosek, "The Fossil Fuel Industry's Push."

25. Schumann, *Disciplining Coal Resistance.*

26. Estes, *Our History Is the Future.*

27. Brock and Dunlap, "Normalising Corporate Counterinsurgency."

28. Healy and Barry, "Politicizing Energy Justice"; Ayling and Gunningham, "Non-State Governance and Climate Policy."

29. Green, "Anti-Fossil Fuel Norms."

30. McGregor, "Direct Climate Action."

31. Heynen, Kaika, and Swyngedouw, *In the Nature of Cities.*

32. Swyngedouw, "Circulations and Metabolisms."

33. Cronon, *Nature's Metropolis.*

34. Arboleda, "In the Nature of the Non-City."

35. Delina, *Emancipatory Climate Actions.*

36. Heynen, Kaika, and Swyngedouw, *In the Nature of Cities.*

37. Swyngedouw, "The Antinomies of the Postpolitical City."

38. Ranciere, *Disagreement: Politics and Philosophy.*

39. Kakenmaster, "Articulating Resistance."

40. Mouffe, *The Democratic Paradox.*

41. Malm, *How to Blow Up a Pipeline.*

42. Ernston and Swyngedouw, *Urban Political Ecology*, 17.

43. Lazarus and van Asselt, "Fossil Fuel Supply."

44. Ciplet, Roberts, and Kahn, *Power in a Warming World.*

45. *Sveriges Natur*, Regeringen säger nej till att koppla på fossilgas på.

46. Balanyá and Sabido, *The Great Gas Lock-In.*

47. Fossilgasfällan, "Protest mot GO4LNGs byggstart den 11 februari."

48. Cheon and Urpelainen, *Activism and the Fossil Fuel Industry.*

49. Folk mot Fossilgas, videos.

50. Howarth, "A Bridge to Nowhere."

51. Vandepitte, Vandermoere, and Hustinx, "Civil Anarchizing."

52. Folk mot Fossilgas, "The Biggest Lie in the History of the Planet."

53. Temper et al., "Movements Shaping Climate Futures."

54. Carton, "Carbon Unicorns and Fossil Futures"; Watt, "The Fantasy of Carbon Offsetting."

References

Ayling, Julie, and Neil Gunningham. "Non-state Governance and Climate Policy: The Fossil Fuel Divestment Movement." *Climate Policy* 17, no. 2 (2017): 131–49.

Arboleda, Martin. "In the Nature of the Non-City: Expanded Infrastructural Networks and the Political Ecology of Planetary Urbanization." *Antipode* 48, no. 2 (2016): 233–51.

Balanyá, Belen, and Pascoa Sabido. *The Great Gas Lock-In: Industry Lobbying behind the EU Push for New Gas Infrastructure.* Brussels: Corporate Europe Observatory, 2017. https://corporateeurope.org/sites/default/files/the_great_gas_lock_in_english_.pdf.

Bassey, Nimmo. "Leaving the Oil in the Soil: Communities Connecting to Resist Oil Extraction and Climate Change." *Development Dialogue* 61 (2012): 332–39.

Bathiany, Sebastian, Vasilis Dakos, Marten Scheffer, and Timothy Lenton. "Climate Models Predict Increasing Temperature Variability in Poor Countries." *Science Advances* 4, no. 5 (2018). https://www.science.org/doi/10.1126/sciadv.aar5809.

Brock, Andrea, and Alexander Dunlap. "Normalising Corporate Counterinsurgency: Engineering Consent, Managing Resistance, and Greening Destruction around the Hambach Coal Mine and Beyond." *Political Geography* 62 (2018): 33–47.

Carton, Wim. "Carbon Unicorns and Fossil Futures: Whose Emission Reduction Pathways Is the IPCC Performing?" In *Has It Come to This? The Promises and Perils of Geoengineering on the Brink*, edited by J. P. Sapinski, Holly Jean Buck, and Andreas Malm, 34–49. New Brunswick, NJ: Rutgers University Press, 2020.

Chatterton, Paul, David Featherstone, and Paul Routledge. "Articulating Climate Justice in Copenhagen: Antagonism, the Commons, and Solidarity." *Antipode* 45, no. 3 (2013): 602–20.

Cheon, Andrew, and Johannes Urpelainen. *Activism and the Fossil Fuel Industry*. London: Routledge, 2018.

Ciplet, David, J. Timmons Roberts, and Mizan R. Khan. *Power in a Warming World: The New Global Politics of Climate Change and the Remaking of Environmental Inequality*. Earth System Governance. Cambridge, MA: MIT Press, 2015.

Cronon, William. *Nature's Metropolis: Chicago and the Great West*. 1991. Repr., New York: W. W. Norton, 2009.

Delina, Lawrence. *Emancipatory Climate Actions: Strategies from Histories*. Cham, Switzerland: Palgrave, 2019.

Environmental Justice Atlas. "Blockadia: Keep Fossil Fuels in the Ground." https://ejatlas.org/featured/blockadia (accessed February 23, 2020).

Ernstson, Henrik, and Erik Swyngedouw, eds. *Urban Political Ecology in the Anthropo-Obscene: Interruptions and Possibilities*. London: Routledge, 2018.

Estes, Nick. *Our History Is the Future: Standing Rock versus the Dakota Access Pipeline, and the Long Tradition of Indigenous Resistance*. London: Verso, 2019.

Folk mot Fossilgas. "The Biggest Lie in the History of the Planet." Facebook, November 27, 2019. https://www.facebook.com/471287340341823/videos/510251913172153.

Folk mot Fossilgas. Videos. Facebook. https://www.facebook.com/watch/folkmotfossilgas/.

Fossilgasfällan. "Protest mot GO4LNGs byggstart den 11 februari." Facebook, February 11, 2018. https://www.facebook.com/Fossilgasfallan/photos/?tab=album&album_id=384223848711569.

Green, Fergus. "Anti-Fossil Fuel Norms." *Climatic Change* 150, no. 1 (2018): 103–16.

Healy, Noel, and John Barry. "Politicizing Energy Justice and Energy System Transitions: Fossil Fuel Divestment and a 'Just Transition.'" *Energy Policy* 108 (2017): 451–59.

Heynen, Nick, Maria Kaika, and Erik Swyngedouw, eds. *In the Nature of Cities: Urban Political Ecology and the Politics of Urban Metabolism*. Vol. 3. London: Taylor and Francis, 2006.

Höök, Mikael, and Xu Tang. "Depletion of Fossil Fuels and Anthropogenic Climate Change—A Review." *Energy Policy* 52 (2013): 797–809.

Howarth, Robert W. "A Bridge to Nowhere: Methane Emissions and the Greenhouse Gas Footprint of Natural Gas." *Energy Science and Engineering* 2, no. 2 (2014): 47–60.

Kakenmaster, William. "Articulating Resistance: Agonism, Radical Democracy, and Climate Change Activism." *Millennium* 47, no. 3 (2019): 373–97.

Kenis, Anneleen, and Mathias Lievens. "Searching for 'the Political' in Environmental Politics." *Environmental Politics* 23, no. 4 (2014): 531–48.

Klein, Naomi. *This Changes Everything: Capitalism vs. the Climate*. New York: Simon and Schuster, 2014.

Lazarus, Michael, Peter Erickson, and Kevin Tempest. "Supply-Side Climate Policy: The Road Less Taken." SEI Working Paper 13, 2015.

Lazarus, Michael, and Harro van Asselt. "Fossil Fuel Supply and Climate Policy: Exploring the Road Less Taken." *Climatic Change* 150, nos. 1–2 (2018): 1–13.

Le Billon, Philippe, and Berit Kristoffersen. "Just Cuts for Fossil Fuels? Supply-side Carbon Constraints and Energy Transition." *Environment and Planning A: Economy and Space* 52, no. 6 (2020): 1072–92.

Malm, Andreas. *How to Blow Up a Pipeline*. London: Verso Books, 2021.

Machin, Amanda. *Negotiating Climate Change: Radical Democracy and the Illusion of Consensus*. London: Zed Books, 2013.

McGregor, Callum. "Direct Climate Action as Public Pedagogy: The Cultural Politics of the Camp for Climate Action." *Environmental Politics* 24, no. 3 (2015): 343–62.

Misutka, P. J., C. K. Coleman, P. Devereaux Jennings, and A. J. Hoffman. "Processes for Retrenching Logics: The Alberta Oil Sands Case, 2008–2011." In *Institutional Logics in Action, Part A*, edited by E. Boxenbaum and M. Lounsbury, 131–63. Bingley, UK: Emerald Group, 2013.

Mouffe, Chantal. *The Democratic Paradox*. London: Verso, 2000.

Nace, Ted, L. Plant, and James Browning. "The New Gas Boom: Tracking Global LNG Infrastructure." Global Energy Monitor, 2019. https://globalenergymonitor.org/report/the-new-gas-boom/.

Nosek, Grace. "The Fossil Fuel Industry's Push to Target Climate Protesters in the U.S." *Pace Environmental Law (PELR) Review* 38, no. 1 (2020). https://ssrn.com/abstract=3769485.

Piggot, Georgia. "The Influence of Social Movements on Policies that Constrain Fossil Fuel Supply." *Climate Policy* 18, no. 7 (2017): 942–54.

Rancière, Jacques. *Disagreement: Politics and Philosophy*. Minneapolis: University of Minnesota Press, 1999.

Rosewarne, Steve, James Goodman, and Rebecca Pearse. *Climate Action Upsurge: The Ethnography of Climate Movement Politics*. London: Routledge, 2013.

Schumann, Vera. "Disciplining Coal Resistance." Master's thesis, Uppsala University, 2019.

SEI, IISD, ODI, Climate Analytics, CICERO, and UNEP. "The Production Gap: The Discrepancy between Countries' Planned Fossil Fuel Production and Global Production Levels Consistent with Limiting Warming to 1." 2019. https://www.iisd.org/publications/production-gap-discrepancy-between-countries-planned-fossil-fuel-production-and-global.

Sippel, Sebastian, Nicolai Meinshausen, Erich Fischer, Eniko Székely, and Reto Knutti. "Climate Change Now Detectable from Any Single Day of Weather at Global Scale." *Nature Climate Change* 10, no.1 (2020): 35–41.

Sveriges Natur. "Regeringen säger nej till att koppla på fossilgas på stamnätet" ("The Government Says No to Switching to Fossil Gas on the Main Grid"). October 10, 2019. https://www.sverigesnatur.org/aktuellt/tillstandet-for-gasterminalen-i-goteborg-avslas-av-regeringen/.

Swyngedouw, Erik. "Circulations and Metabolisms: (Hybrid) Natures and (Cyborg) Cities." *Science as Culture* 15, no. 2 (2006): 105–21.

Swyngedouw, Erik. "The Antinomies of the Postpolitical City: In Search of a Democratic Politics of Environmental Production." *International Journal of Urban and Regional Research* 33, no.3 (2009): 601–20.

Swyngedouw, Erik. "Apocalypse Forever?" *Theory, Culture, and Society* 27, nos. 2–3 (2010): 213–32.

Temper, Leah, Sofia Avila, Daniela Del Bene, Jennifer Gobby, Nicolas Kosoy, Philippe Le Billon, Joan Martinez-Alier, et al. "Movements Shaping Climate Futures: A Systematic Mapping of Protests against Fossil Fuel and Low-Carbon Energy Projects." *Environmental Research Letters* 15, no. 12 (2020): 123004.

Tong, Dan, Qiang Zhang, Yixuan Zheng, Ken Caldeira, Christine Shearer, Chao-peng Hong, Yue Qin, and Steven J. Davis. "Committed Emissions from Existing Energy Infrastructure Jeopardize 1.5 C Climate Target." *Nature* 572, no. 7769 (2019): 373–77.

Vandepitte, Ewoud, Frederic Vandermoere, and Leslie Hustinx. "Civil Anarchizing for the Common Good: Culturally Patterned Politics of Legitimacy in the Climate Justice Movement." *Voluntas: International Journal of Voluntary and Nonprofit Organizations* 30, no. 2 (2019): 327–41.

Walker, Gordon. *Environmental Justice: Concepts, Evidence, and Politics.* London: Routledge, 2012.

Watt, Robert. "The Fantasy of Carbon Offsetting." *Environmental Politics* 30, no. 7 (2021): 1069–88. https://doi.org/10.1080/09644016.2021.1877063.

Climate Insurgency between Academia and Activism

An Interview with David N. Pellow

Marco Armiero and Salvatore Paolo De Rosa

David N. Pellow is the Dehlsen Chair and Program Chair of Environmental Studies and director of the Global Environmental Justice Project at the University of California, Santa Barbara. He is a committed radical scholar working with students and communities to foster social and environmental justice. In recent years, David Pellow has worked on two projects: one on the environmental injustice in the prison system and the other on a bottom-up Green New Deal with the subaltern communities of central California.

Marco Armiero and Salvatore Paolo De Rosa: *With the expression "urban climate insurgency," we refer to the ensemble of radical practices that reject the climate consensus while fostering an antagonist agenda about climate change. Do you see a potential or existing convergence between radical movements such as Black Lives Matter (BLM) and climate justice?*

David N. Pellow: I would like to respond to this question by referring to the critical work that the Central Coast Climate Justice Network (C3JN) is undertaking. C3JN is a multiracial network of social justice and environmental organizations and leaders committed to a climate movement that advances social, economic, and environmental justice for California's Central Coast communities [the region just north of the City of Los Angeles, including Ventura, Santa Barbara, and San Luis Obispo Counties]. There are members of this network who have long been active in supporting both the cause of climate justice and Black Lives Matter, so in June of 2020, when C3JN penned "Letter of Support and Solidarity re: Black Lives Matter,"

Social Text 150 · Vol. 40, No. 1 · March 2022
DOI 10.1215/01642472-9495174

that was only the latest effort to articulate a convergence. That letter began as follows:

> There is no climate justice without racial justice. We, the Central Coast Climate Justice Network (C3JN) affirm that the lives, dreams, guidance, wisdom, lived experiences and futures of Black people and peoples of African descent in the U.S. and the world matter. We write these words with great conviction and solidarity: Black Lives Matter. They matter to Black communities and they matter to all of us because there can be no freedom while any of us is oppressed.[1]

In the US under the reign of racial terror associated with centuries of white supremacy and the most recent amplification of that brutal system under the Trump regime, to even utter the words "Black Lives Matter" can be judged to be an act of sedition and an embrace of "terrorism."[2] Trump and much of the political and media establishment could not bring themselves to even perfunctorily utter these words without expressing open contempt, ridicule, and invoking threats of state-sanctioned racist violence directed against those who would dare to support the right of Black people to exist, let alone thrive. Thus for climate justice organizers to do so and to explicitly and cogently articulate support of the aims of BLM is bold and significant. That move is also a welcome step forward considering the long history of exclusions, erasures, and anti-Black racism that has characterized the traditionally white, middle-class environmental movement. Going further, the C3JN continued in its statement of solidarity and offered a critique of the narrative of "existential threat" we hear so much in the mainstream climate movement regarding the climate crisis:

> As climate justice activists, it is imperative that we bring the same level of urgency to the struggle for racial justice that we bring to our efforts to address climate change. Much of the climate emergency discourse from environmentalists assumes that climate disruption is the first time that humans have been threatened with an existential crisis. Nothing could be further from the truth. As the experiences with genocide, colonialism, enslavement, and other forms of state and institutional violence that Black communities have endured amply indicates, we must acknowledge that our Black brothers and sisters have always had to fight for their existence.[3]

This portion of the statement really speaks for itself, but it must also be noted that it is a powerful correction to the "emergency" narrative that implicitly centers white lives and experiences and ignores the centuries of racist violence directed at peoples of African descent. And since Black people have endured and survived such pain and brutality for generations, we are urging the world to look to those communities for leadership and wisdom with respect to how one might push through times of crisis and remain intact. This is a point that many others have made with respect to the depth of experience and knowledge that frontline, fence line, and BIPOC communities can offer the world in our collective efforts to address and confront climate disruption.[4]

C3JN's letter of support also endorses the national demands of the Movement for Black Lives while explicitly calling for allyship with Indigenous peoples as well, underscoring the exceedingly important point that just focusing on the struggles and aspirations of any single population will limit our overall efforts, since this is and must be a "big tent" movement that is opposed to any form of domination wherever and whenever it rears its head. Going a necessary step further, C3JN is integrating many of the demands of the Movement for Black Lives into its proposed Green New Deal framework for the region. That is all to say that this is but one of many clear indications and examples of public support for—if not a convergence among—BLM climate justice movements.

You have been very active in bridging community activism and academic work around issues of environmental and climate justice. Can you tell us something about your experience? Did it work? What kind of resistance did you encounter from the academic side and also from the community?

I have been extremely fortunate to have always had a wonderful group of students, faculty, staff, and community activists as collaborators and colleagues. One of my earliest projects involved supporting the launch of the International Campaign for Responsible Technology, a group of environmental justice and labor rights activists and scholars around the world who have succeeded in building and supporting movements to advocate for workers in electronics/IT industries and the fence-line communities impacted by that sector's antilabor practices and its manufacture and use of an enormous volume of toxic materials. The ICRT has successfully pushed some of the world's largest IT companies, along with the European Union and many other governments and university systems, to adopt its proposals for environmental stewardship and labor protection. While there are many drawbacks and loopholes in these efforts, they are important for demonstrating that grassroots movements—supported by scholars and the research we produce—can produce meaningful change on multiple scales in the service of environmental justice, climate justice, and labor justice. But this movement is also urgently needed because it reveals that levers and targets of change can include and extend beyond the nation state. For example, the Silicon Valley Toxics Coalition—a lead organization in the founding of the ICRT—has long used the practice of developing "score cards" to publicly name and shame corporate environmental offenders and move them toward greater accountability and behavioral changes. And they have succeeded in doing so numerous times, particularly with corporations that have little incentive to do otherwise since government regulators and existing legislation do not require such changes. While SVTC and ICRT certainly invest energy in pushing for states to change and enact policy, this is but one approach to environmental justice they embrace.

I am also privileged to have worked with the Ecuadorian NGO DECOIN (Defensa y Conservacion Ecologica de Intag/Intag Defense and Ecological Conservation group) to develop a written guide to support activists seeking to defend their communities against "extractivist" practices. DECOIN is based in the Intag cloud forest region of northwestern Ecuador, which is home to numerous low-income, rural communities and dozens of endangered nonhuman animal and plant species. Prior to our collaboration, DECOIN had successfully worked to push back and even expel mining companies that had begun devastating parts of their communities and critical ecological habitats. At the time, there were few written resources for activists wishing to address these challenges, so we produced one that is downloadable for free from the Cultural Survival organization's website.[5] This guide was translated into several languages and has been used by advocates in many countries. I was proud of my involvement in this work but was not prepared for the threat of repression that my colleagues in Ecuador soon faced. That nation's president at the time, Rafael Correa, took to national television and denounced our guide and us—the authors—during a press conference. He referred to us as threats to national security and, for a while, it was unclear what would happen to our colleagues on the ground there. Ultimately, they were allowed to continue their work but under a heightened sense of (in)security and with the possibility of increased state repression. I would say that while I tend to be concerned about the backlash or lack of support for my activist-scholarship work from *academic* colleagues and institutions, as a North American scholar with tenure, I rarely have had to worry about risks to my physical safety in response to my writings and activism. That is a luxury that many of my colleagues in the Global South do not have, and I will keep that at the top of my consciousness from this day forward.

Lately, you have been reflecting on the agency of the state in environmental injustices, uncovering the contradictions of demanding justice to the very actor who has produced injustice in the first place. If climate change is not a mistake in the system but is the way racial capitalism works, what does this imply for the climate justice movement?

While there is increasing emphasis on confronting capitalism and building support for anticapitalist theory and politics among scholars and movements focused on climate and environmental justice, I find it curious that there is only a nascent engagement with equally rigorous and strident critiques of state power from those same quarters. If I can say, and I certainly have, that capitalism requires calculated brutality and is fundamentally incompatible with the goals of improving and sustaining human and environmental health, then I can most definitely also say the same for those forms of governance that uphold and constitute the modern nation-state. I am certainly sympathetic to those observers and scholars who rightly point to critical progress we have made on human and civil rights and specific environmental protection efforts as a result of urging governments to deliver on those demands over the years. Those arguments are legitimate for particular places and moments in time, but, unfortunately, they conveniently disregard three uncomfortable truths: (1) social inequality within and between nation-states is at its most extreme ever in the current era, (2) the enslavement and trafficking of human beings today is far more extensive than has occurred ever before in human history, and (3) anthropogenic climate disruption and the present-day massively destructive consumption of land animals and marine life are unparalleled in the course of our collective histories. Given this indisputable evidence of decline, despoilation, and the sixth mass extinction during the reign of modern nation-states, what indicators or data could possibly give any climate or environmental activist or scholar even a shred of confidence that the same system could somehow reverse course, undo these harms, and shed its skin to morph into something entirely different? And yet the climate justice, environmental justice, and racial justice movements continue to press forth with this assumption. Each of these formations—and the scholarly literatures that parallel them—fervently support the notion that we can and will secure some semblance of justice and equitable futures through the mechanisms set up precisely to deny us those things. I understand the logic of looking to structures and institutions that contain such enormous power and potential to confront the myriad of massive and intersecting crises we are enduring, but I think we ought to be far more cautious and imaginative about this quest.

Let us consider the so-called climate insurgency. Jeremy Brecher's trilogy[6] on this topic is, in many ways, compelling and inspiring and reflects the widely held view among environmentalists that we can address the climate crisis through existing institutional and legal frameworks. I certainly agree that we can work to *slow the rate* of damage and destruction caused by state and corporate global socioecological violence, but there is little evidence to suggest that we can reverse these trends using these "master's tools." However, I take a pragmatic approach to this challenge by borrowing from scholars like James Scott and Arturo Escobar, and I agree that *for now* we can occasionally work through the state to achieve gains and to blunt the worst of the traumatic consequences of the system, but I see no reason to imagine that this will be a sustainable or equitable long-term strategy, and scholars like Laura Pulido have concurred in far more eloquent language than I can offer up here.

Even during the recent resurgence of BLM mobilizations around the US and the world, that movement continues to articulate a firm reliance on the state. One might think this is not the case, considering the popularized call to "defund the police," which seems to have a heady anarchist aura about it. However, even this seemingly far-reaching proposal is squarely state-centric because activists are not calling for the abolition of policing or the state apparatus that supports it; they are simply pushing for a reallocation of some state resources from policing to other urgently needed sites—health care, education, housing, etc. I certainly agree that we need more resources for those critical goals, but I am concerned that BLM is firmly of the view that the same state that is extinguishing the lives and dreams of Black people can be commandeered to do the opposite for a sustained period of time.

What I *am* in favor of and excited about is the extraordinary rise of mutual aid networks around the world, mobilizing to deliver critical resources to marginalized and vulnerable populations. While I could point to any number of such efforts around the globe, I will again quote from the California Central Coast Climate Justice Network's BLM solidarity statement as an example of this transformative work:

> In recent weeks, many of us have been working to provide mutual aid to people in need of food and basic services in the absence of a functioning federal government and health care system; protesting in the streets for racial justice in a nation that perpetuates unrestrained violence against many of its citizens simply because of their racial-ethnic heritage; and sending support to communities around the U.S. and the world during this time of great need and deprivation. We will continue that work, taking our lead from Black community activists whose voices and knowledge must be centered in this struggle.[7]

Continuing with the theme of state engagement, one of my favorite and least favorite names of a climate movement group is Extinction Rebellion. It is one of my favorite names because it explicitly underscores what is at stake for so many of us humans and our other-than-human relations—extinction versus survival—and that what we need is in fact a rebellion to ensure our continued existence on earth. But it is my least favorite name of a climate movement endeavor precisely because it is *mis*named. As much as I applaud and support their work of raising the alarm, pushing institutions to be more accountable to the realities of climate disruption, and making these concerns more visible to larger audiences and potential supporters, Extinction Rebellion is anything but a *rebellion*. It is, rather, a very important protest movement. First, speaking to the question of the role of the state, Extinction Rebellion is wholly reliant on a state-centric set of tactics and strategies. Their top three demands[8] each begins with the word "Government." Namely, they are: (1) "governments must tell the truth," (2) "governments must act now," and (3) "governments must create." There is little room in this framework for understanding how ordinary people can lead and make change in the absence of the overwhelming and inherently authoritarian presence of nation-states, and that gives me great pause. Second, as an African American, I must say that when I heard the word "rebellion," I think of my ancestors who, during the era of formal chattel enslavement in the Americas, rebelled not by asking the slave masters to provide them with better working conditions while ensuring the maintenance and survival of the system of human bondage. Rebellion in that context meant a vision and practice of *overthrowing* the system and liberating people from its intrinsically oppressive functions. Extinction Rebellion might take a page from that history book, because while it is, as I say, an important protest movement, the extent of its actions, strategy, and vision are to clamor for change and reform from the very states and corporations that drive and profit from socioecological and climate harm.

To conclude on a more positive and forward-leaning note, I am delighted that a number of scholars and activists are now articulating the ideas of abolitionist climate and environmental justice—a multi-issue politics and analysis that addresses historical harms through a decolonial and anticapitalist framework while investing in an ethic of care for those populations most affected by environmental and climate injustices. In that vein, I propose fusing the insights of abolitionists and multispecies justice scholars to offer a vision of what I call "multispecies abolition democracy," which I define as those practices, institutions, and structures that enable and facilitate justice for humans and nonhumans in the context of recognizing that since our societies have always been multispecies in character and membership, our systems of decision-making should be as well. And since abolition democracy was a framework intended for humans only, multispecies abolition democracy builds on that inspiring vision and extends and deepens it so as to allow for all beings and things to be recognized as members of our societies and that collaboration rather than exclusion and domination are practices and ethics that will strengthen our communities in ways that are truly and "deeply intersectional" because they involve and include the vulnerable and the privileged within and across the human and species boundaries. When I use terms like "decision-making" and "democracy," I eschew the state as the most desirable embodiment of these practices; rather, I see these terms as signaling a set of values, practices, processes, and actions rather than primarily an institutional or organization form. This is a vision that will require commitment and labor from advocates across many social movements who have thus far generally taken modest steps in that direction. But I believe there are important opportunities and a clear and urgent rationale for pursuing that bold and courageous project.

Marco Armiero is research director at the Institute for Studies on the Mediterranean, CNR (Italian National Research Council) and director of the Environmental Humanities Laboratory, KTH–Royal Institute of Technology, Sweden. He is the author of the book *Wasteocene: Stories from the Global Dump* (2021).

Salvatore Paolo De Rosa is a postdoctoral researcher at Lund University Centre for Sustainability Studies, Sweden, and an activist for climate justice. His background is in anthropology and geography while his current research focuses on the political ecology of climate movements.

Notes

1. Central Coast Climate Justice Network (C3JN) Letter.
2. See, for example, BLM cofounder Patrisse Khan-Cullors's book *When They Call You a Terrorist: A Black Lives Matter Memoir.*
3. Central Coast Climate Justice Network (C3JN) Letter.
4. See, for example, Sarah Krakoff's article in *Environmental Justice*, "Radical Adaptation, Justice, and American Indian Nations."
5. Zorrilla et al., *Protecting Your Community.*
6. Brecher, Climate Insurgency; Brecher, *Climate Solidarity*; Brecher, *Against Doom.*
7. Central Coast Climate Justice Network (C3JN) Letter.
8. Extinction Rebellion, "Our Demands."

References

Breceher, Jeremy. *Against Doom: A Climate Insurgency Manual.* Oakland, CA: PM, 2017.

Brecher, Jeremy. *Climate Insurgency: A Strategy for Survival.* West Cornwall, CT: LNS/Stone Soup, 2017.

Brecher, Jeremy. *Climate Solidarity: Workers vs. Warming.* West Cornwall, CT: LNS/Stone Soup, 2017.

Central Coast Climate Justice Network (C3JN) Letter of Support and Solidarity re: Black Lives Matter, June 2020. https://resource.cecsb.org/c3jn-letter-support-solidarity-black-lives-matter/.

Extinction Rebellion. "Our Demands." https://extinctionrebellion.uk/the-truth/demands/.

Khan-Cullors, Patrice, and Bandele Asha. *When They Call You a Terrorist: A Black Lives Matter Memoir.* New York: St. Martin's, 2018.

Krakoff, Sarah. "Radical Adaptation, Justice, and American Indian Nations." *Environmental Justice* 4, no. 4 (2011): 207–12.

Zorrilla, Carlos, Buck Arden, Palmer Paula, and Pellow David. *Protecting Your Community against Mining Companies and Other Extractive Industries: A Guide for Community Organizers.* 2009. https://www.culturalsurvival.org/news/protecting-your-community-against-mining-companies-and-other-extractive-industries-guide.